洛带会馆 客地原乡

Luodai Guild Halls Hometowns in A Foreign Land

主　编　张利频　曾　列

执行主编　张　珂

四川出版集团·四川美术出版社
SICHUAN PUBLISHING GROUP
SICHUAN FINE ARTS PUBLISHING HOUSE

序言

洛带会馆
历史空间延续的历史生命

会馆者，顾名思义，乃是"聚会之馆"，而"馆"，古人的解释是"馆舍也。凡事之宾客，馆焉"，那么，什么样的"馆"才是宾客聚会的"会馆"呢？应该说会馆的雏形源自汉代的郡邸，那是汉时长安专为同乡人提供驻足栖歇的一种旅馆，借住其中，虽远在他乡，却乡音缭绕，吃到故乡可口的饭菜。

学界认为，明永乐年间安徽芜湖人在北京设置的芜湖会馆，是中国历史真正意义上的会馆。其历史背景是，明永乐十三年（1415年），永乐皇帝朱棣将在南京举行的科举考试移到北京，一时间，北京城里生员荟萃，一些为官之人便邀集同乡士绅商人，建造馆舍，在服务官绅的同时，也为本乡学子提供一个公车谒选的寄宿之地。这样的会馆，又称"试馆"，为文人会馆。随着明清时代商业的发达，散布于各地水路要冲、繁华商埠的会馆，则多为为商旅而设的商业会馆或同业会馆，它们或以同乡商人籍贯的地名冠名，如山陕会馆；或以行业冠名，如盐业会馆。

因地处川渝古商道上，成都近郊的洛带古镇，是我国现存会馆最多的乡镇。从上场口到下街，至今依然弥足珍贵地排列着川北会馆、广东会馆、江西会馆和湖广会馆。这些会馆建筑布局考究，建造精美。从会馆的功用来看，洛带的几大会馆都属移民会馆。所谓移民会馆，地方志的阐释是，"清初各省移民来填川者，暨本省遗民，互以乡谊连名建庙，祀以故地名神，以资会合者，称为会馆。"至于其功能，则是"迎麻神，聚嘉会，襄义举，笃乡情。"可见移民会馆的本意是"庙"，是祭祀乡梓神祇的地方，是赈济贫困同乡，促进同籍情谊的族群聚合地。

明末清初，四川因战乱、灾荒、瘟疫等，曾经丰腴的天府之国，满目疮痍，人口锐减，一片凋敝。《清史稿·食货志》称当时的四川"孑遗者百无一二"，清初张烺的《烬余录》也说"蜀民至是殆尽矣"。在这一历史背景下，清政府采取了系列措施，动员湖广、岭南等省的居民前往四川拓荒生息——这就是历史上著名的"湖广填四川"。到清末，以省会成都为例，按照《成都通览》的记述，"现今之成都人，原籍皆外省人"。时有《锦城竹枝词》曾这样描述："大姨嫁陕二姨苏，大嫂江西二嫂湖。戚友初逢问原籍，现无十世老成都"，形象地说明了当时成都的人口构成。

身在异乡为异客，各地移民涌向四川后，为适应新的生存环境，"人多力量大"的同乡之情，自然就代替了势单力薄的亲友之情，各种以"乡梓之情"为基础的同籍移民组织孕育而生。在这些组织的张罗下，寄望于联络同籍情谊、将"故乡文化"移植过来的会馆建筑，也就在移民们客居之乡相继建造起来，并成为一方地域的地标性建筑。

在"湖广填四川"滚滚移民浪潮中，先期来到洛带的移民是湖广人，之后是粤赣籍的客家人，所以洛带先有湖广会馆，之后才有了广东会馆和江西会馆。而据考证，在湖广人到达洛带之前，还有一批陕西、山西籍的"秦晋人"来到洛带并修建了会馆秦晋宫，不过早已消失无存。秦晋宫的消失，充分显示了当时的洛带移民聚居区，一种文化的流变和衰亡，和对另一强势文化的认同。相比较而言，在洛带四大会馆中，以广东会馆最气势轩昂，以江西会馆最精巧别致，这也从侧面说明了这一地域客家文化的兴盛发达。至于上场口的川北会馆，那是2000年因城市建设需要，从成都卧龙桥街异地原貌搬迁至此的，川北会馆的异地原貌重建，使洛带作为"会馆之乡"更加名至实归。

但客家人初到四川的处境可谓举步维艰，广东会馆有两柱楹联就谈到，客家人一度"惨淡经营"，但"努力缔造"，最终"华（成都华阳）简（四川简阳）俱成桑梓地，乡音无改，新增天府冠裳"，从中展示出客家人抱聚成团、共克时艰、勇于开拓的族群意识和乐观精神；同时也反映出他们视他乡为故乡，建设美好四川的良好愿望。在其时千里凋敝、百废待兴的四川大地，包括客家人在内的移民们没有沉沦，除却总总，会馆功不可没。原因很简单，在会馆里，他们可以互诉乡愁，共享欢欣，互助互爱，同舟共济，所以，一定程度上讲，会馆就是移民们在异土的原乡，是他们源源不断获取力量的精神华表和心灵归依的地方。

今天，"乡音无改"的洛带，被学界称为中国内陆"最后的客家王国"。依托会馆建筑群，和当地丰厚的客家文化资源，移民会馆荟萃的洛带正成为中外游客的热游之地，而饱含中国传统礼俗文化、道德规范的会馆文化也因此得到更加广阔地传播和弘扬。或许可以这样来看待，作为一种珍贵的物质文化遗产，会馆不仅仅是一部由土木材料书写的史学民俗巨著，更是我们触之有温、溯源根脉的精神家园；是传承文脉，历史空间延续的历史生命。

刘晓峰 2013年6月于北京

刘晓峰 全国政协副主席

Luodai Guild halls
Lives of History in A Historic Space

Guild hall, or huiguan in Chinese, as its name implies, is "a hall for guild". "Guan", according to ancient Chinese people, meant"a house offered to guests". Which kind of a "house" can be offered to guests as a "guild hall"? It was said that the rudiment of a guild hall originated from offices of prefectures in Chang' an, the capital during the Han Dynasty. It was a kind of inn in Chang' an offered to people from the same native place to rest in, where the same-dialect groups could obtain food in their hometown style while being away from home.

Generally, academic experts think that the Wuhu Guild Hall in Beijing, set up by people from Wuhu, Anhui during Yongle Emperor of the Ming Dynasty, was the first real guild hall in Chinese history. If true, this would be against a backdrop to the 13th year Yongle Emperor of the Ming Dynasty (1415). Yongle Emperor, Zhu Di, moved the imperial competitive examination from Nanjing to Beijing, led to an influx of students into the new site. Some officials in Beijing invited gentries and businessmen from the same locale to establish halls as a lodge for students from their hometown during their examination and interviews in Beijing. Such guild hall, named as "examination hall", was a hall for intellectuals. During the Ming and Qing Dynasty, guild halls in water hubs flourished as trading ports and halls for traveling businesspeople. They were named according to the name of their hometowns or the industries of the occupants such as Shanxi and Shaanxi Guild Hall or Salt Industry Guild Hall.

As an ancient town located along the Sichuan-Chongqing trade passageway and in the outskirts of Chengdu, Luodai has the largest number of guild halls among Chinese villages and townships; North Sichuan Guild Hall, Guangdong Guild Hall, Jiangxi Guild Hall, and Huguang Guild Hall to name a few. Their architectures are exquisite construction. In terms of functions, these guild halls in Luodai are immigrant guild halls. According to local chronicles, the immigrant guild hall means that "at the beginning of the Qing Dynasty, immigrants from other provinces came to Sichuan and established temples based on their provincialism, where they gathered and enshrined famous gods from their hometown. Those temples were also called guild halls." Their functions included "hosting Ma God, holding feast, cooperation in chivalrous deed, and being loyal to provincialism." The original meanings of immigrant halls are "temples", a gathering cluster for people from the same hometown to cherish the memory of their native place, worship deity, succor needy fellows, and build friendship among them.

At the turning point of the Ming and Qing Dynasties, due to chaos caused by wars, disasters, and pestilences, the fertile land of Abundance became destitute and devastated with a sharply declined population. According to History of the Qing Dynasty: Economic Chronicles, survivors took up an extremely small part of the people in Sichuan at the time. The Records for Survivors, written by Zhang Lang, a writer living during the beginning of the Qing Dynasty also mentioned that "Sichuan people disappeared until now". Against such a backdrop, the government of the Qing Dynasty took a series of measures to encourage people living in Hubei, Hunan, Guangdong, and Guangxi to reclaim wasteland and move to Sichuan. The migration wave was known as "Huguang Filling Sichuan". Up to the end of the Qing Dynasty, according to Encyclopedia on Chengdu, "many people in Chengdu came from other provinces". It was also mentioned in Jincheng Zhuzhici (Jincheng is another name of Chengdu and Zhuzhici refers to ancient folk songs with love as their main theme): "The eldest aunt was married to a person in Shaanxi and the second eldest aunt in Jiangsu; the eldest sister-in-law is from Jiangxi and the second eldest sister-in-law from Hunan or Hubei. When relatives or friends ask each other where they are from during their first meeting, there are even no native Chengdu people whose ancestors ten generations ago were truly native". Such a description lively reflects population composition in Chengdu at the time.

As strangers in a strange place, immigrants from other provinces flooded into Sichuan paid more attention to the relationship among people from the same province instead of among relatives and friends because "more hands provide greater strength and the former has a large scale than the latter". As a result, various organizations were established to serve this chain migration movement on the basis of "feelings for hometown". Due to the care of those organizations, guild hall architectures functioned to make friendly contacts among people from the same province and cultures in their hometown were constructed successively in immigrants' new residence and become a landmark in the area.

The pioneers of immigrant wave to Luodai were people from Hunan and Hubei, followed by the Hakka people from Guangdong and Jiangxi. This is why Huguang Guild Hall was established in Luodai prior to Guangdong Guild Hall and Jiangxi Guild Hall. According to researches, before the arrival of people from Hunan and Hubei, people from "Qinjin (refers to Shaanxi and Shanxi)" came to Luodai and constructed Qinjin Guild Hall, but Qinjin Guild Hall disappeared much earlier. The disappearance of Qinjin Guild Hall shows a cultural change and decline in gathering place of immigrants in Luodai and an emergence of a stronger mainstream culture. Among the four major guild halls in Luodai, Guangdong Guild Hall has a dignified style, and that of Jiangxi exquisiteness and uniqueness, which shows the richness of Hakka culture in this area. As to North Sichuan Guild Hall at Shangchangkou, it was moved from its original appearance on Wolongqiao Street in Chengdu in 2000 to make room for a municipal construction. The reconstruction of North Sichuan Guild Hall followed its original appearance in a new place and this made Luodai became more deserving to the name of "A Township for Guild Hall".

In the beginning, the living condition of the Hakka people in Sichuan was extremely difficult. The couplet hung on the columns of Guangdong Guild Hall mentioned that Hakka people once lived against difficulties, but through "painstaking efforts", they eventually "transformed Hua (short for Huayang) and Jian (short for Jianyang) both into a native place for them, where they still spoke their original dialect. This added a new civilization to Chengdu". It shows that Hakka people overcame difficulties as a united one and dared to forge ahead with their optimistic attitude. This shows their good wishes to make an alien land their hometown. Even facing destitution and devastation, Sichuan as well as Hakka people still used guild halls to make great contributions besides a variety of other driving forces. The reason is simple: in a guild hall, people can tell their feelings to one another, share happiness, and help each other to overcome distress. Therefore, to some degree, a guild is a native place for people in an alien land, where they can reenergize and find a home for their souls.

Today, Luodai is called as the "last kingdom of Hakka culture" by Chinese academia. Based on its guild hall architectures, a rich Hakka culture, and a strong Hakka dialect, Luodai, a cluster of immigrants' guild hall, becomes a hot destination among tourists at home and abroad. It is a place where guild hall culture of Chinese traditional etiquette and customs culture blend in a beautiful landscape. We may think that as a precious material cultural heritage, guild hall is not only a great historical and folk-custom work consisting of construction materials, but also a warm spirit home, where people can trace their source and root. It is a cultural inheritance as a life of history in a historical space.

By **Liu Xiaofeng**
June 2013, Beijing

Liu Xiaofeng, vice chairman of Chinese People's Political Consultative Conference

Ancient Luodai Town, lying in the eastern suburb of Chengdu, is a Hakka town with the most number of existing guild halls in China. Those Guild Halls are the products of the campaign "Huguang Filling Sichuan", one of the largest waves of migration in Chinese history, and mirror the social situation of Sichuan during the past few years. As time goes by, changes are results of the vicissitudes of time. However, the traditional virtues such as maintaining harmony in the neighborhood, showing solicitude for children and orphans, and making honest and good faithful efforts are still advocated and praised by the locals living in the land. A range of cultures relating to village life that coexist with the guild halls are passed down from generation to generation, forming a part of life. The historical charm and cultural profoundness reflected by the buildings of the guild halls are always touching and fascinating nowadays.

成都东郊客家古镇洛带,是中国现存会馆最多的场镇。那些弥足珍贵的会馆,是中国历史上最大规模移民浪潮之一——"湖广填四川"的产物,是晚近四川社会的一面镜子。时光流逝,岁月沧桑,会馆曾经承载的乡谊襄助、恤幼悯孤、敦厚信义等传统美德,依然为这片土地所崇尚赞咏;依会馆而存的种种乡村活态文化,也依然触之有温,血脉相袭;会馆建筑今天展现出来的历史韵味与悠远之美,更是那样地打动人心,令人醉迷。

目 录

序　言
洛带会馆
历史空间延续的历史生命 …… 002

概　述
从汉代"郡邸"到洛带会馆 …… 010

第一章
洛带古镇与洛带会馆 …… 025

第二章
湖广移民与湖广会馆 …… 035

第三章
赣南客家移民与江西会馆 …… 055

第四章
粤籍客家移民与广东会馆 …… 075

第五章
川北会馆
晚清四川会馆的典范 …… 095

第六章
会馆祀奉
乡情联络的纽带 …… 115

第七章
会馆楹联
解读移民心灵密码的钥匙 …… 129

第八章
会馆戏楼
川剧五腔共和的摇篮 …… 139

CONTENTS

PREFACE   /004
Luodai Guild Halls
Lives of History in A Historic Space

SUMMARY   /015
From "Official Mansions" in Han Dynasty
to Luodai Guild Halls

CHAPTER 1   /027
Ancient Luodai Town
and Luodai Guild Halls

CHAPTER 2   /037
Migrants from Huguang
and Huguang Guild Hall

CHAPTER 3   /057
Hakka Migrants from South Jiangxi
and Jiangxi Guild Hall

CHAPTER 4   /077
Hakka Migrants from Guangdong
and Guangdong Guild Hall

CHAPTER 5   /097
North Sichuan Guild Hall
A Classic of Sichuan Guild Halls during the Late Qing Dynasty

CHAPTER 6   /118
Worship in Guild Halls
Links Connecting Affections for Hometown

CHAPTER 7   /131
Couplets Hung on the Columns of Guild Halls
The Keys to the Soul of Migrants

CHAPTER 8   /141
Opera Stages in Guild Halls
Where 5 Common Systematic Tunes of Sichuan Opera Blend

从汉代"郡邸"到洛带会馆

概述

对外部人群而言，洛带就是一处移民会馆聚集之地——川北会馆、广东会馆、江西会馆、湖广会馆，从上场口依次而下，在古镇的各个街段，风情洋洋地耸立起风貌有别的建筑景观。

高墙、深院、飞檐、斗拱；厚门、窄窗、天井、戏楼……多少年来，会馆阔朗的建筑空间，缘结南北来往客，堂聚悲欢离合情；而那些精巧的建筑细节里，又该布满怎样不为人知的历史烟云和人文密码呢？

老式"九宫格"的窗上，正悬挂着今夜的明月……异乡异客，乡音乡情，饱经风霜的建筑群落和生活物什，都镀上了一层神秘而熹微的光彩，那是"客而家焉"移民的呼吸，是客地密境漫长岁月沉淀的迷人历史气质。

会馆——异乡里的原乡

古往今来，行止他乡的人们都需要一处能落脚栖息的地方，史料记载，至迟始于周代，一些城池便设立了可供异乡人寄寓的驿馆。到汉时，地方各郡在长安都设有"郡邸"，当同乡官僚远道而来时，乡音缭绕的郡邸可为客人提供故乡可口的饭菜。及至宋代，各个城市里为接待远来客商、学子而开设的"邸店"，比比皆是。同时，这些邸店开始以地域和行业划分客舍，比如北宋开封潘街楼南有家"鹰店"，就"只下贩鹰鹃客"。之后，南宋临安京师之地，又开始出现一些外地人为本籍来杭乡亲谋求公益的同乡组织。

由此而见，国人历来都非常重视乡梓、同业情谊，并以此作为一种情感联络的纽带，以在异乡谋取共同的利益。但严格意义上讲，它们都非会馆——郡邸，无非相当于今天的"驻京办事处"，而"邸店"之类，其功能则更多偏向于今天的宾馆、旅社。

会馆的出现，依据史料发掘，大体可上溯到明永乐年间。明永乐元年（1402年），靖难之役后朱棣夺位登基，成为明王朝的第三代皇帝。一俟登基，鉴于当时北方房患不绝，为巩固帝国边防，朱棣改北平为北京，开始了将京师从南京迁都北京的一系列建设。事实上，北京城建了十五年，直到明永乐十九年(1421年)正月朱棣才正式下达迁都的诏令，但作为"行在"的北京，从明永乐七年（1409年）起，几乎一直都是朱棣办公的地方，这使得不少当朝流寓京官，为便于上朝谒圣，纷纷邀集同乡士绅、商人等，在北京合力集资购买土地，建造馆舍，以为寄宿、聚会之所。

明永乐十三年（1415年），朱棣又将三年一次的全国性科考会试，从南京改在北京举行。一时间，北京城陡增万人之众，而科考之后，尚有大部分未能折桂的学子因乡途遥远，寄宿北京，以待下一届会试，于是便有好心的同籍官绅，邀请他们同住由这些官员士绅集资修建的会馆里。要知道，在科举时代，名登皇榜，是一个地方的荣耀，所以不少省、州府、县也争相在京城修建为赴考生员提供寄宿方便的会馆。但这样的会馆，《重修歙县会馆录》就明确规定，会馆"创立之意，专为公车以及应试京兆而设，其贸易客商自有行寓，不得于会馆居住以及停顿货物，有失义举本意"。所以起初的会馆又称"试馆"，或者"文人会馆"。但它们之间也有些细微的差异，一般说来，"内城馆者，绅是主；外城馆者，公车岁贡是寓。"

与此同时，一些服务于同乡同行的商业会馆也在各地水路要冲、繁华商埠建造起来。商旅辐辏有会馆，与其说这是商人们对文人会馆的一种模仿，毋宁说是对文人会馆排他行为的一种抗争。中国传统社会有"四民"之称——"士、农、工、商"，商人排位最后，可见，会馆之于商人，也是他们以"桑梓之谊"，"群聚而笃"谋求一种社会地位认同的手段。所以，各地商业会馆运作伊始，便制定了严格的"行规"、"章程"、"俗例"等，目的当然是"荟萃商贾，会聚同乡，公议同业，襄助同仁"，利义共举——利者，商贾之津；义者，儒家之要，会馆不仅是融合工商利益与儒家道德思想的社会容器，一定程度讲，会馆的这种与上层社会意识形态靠拢的自我管理，既顺应了当时社会管理的需要，也是其时社会秩序稳定的基石。所以说狭义上的会馆，是一种建筑形态；而广义上的会馆，是一种社会组织。

会馆由来都是以地缘为联系纽带的，异乡异客，但乡音乡情，所谓"敦亲睦之谊，叙桑梓之乐，虽异地宛若同乡"，所以，会馆实则就是流寓之人在异乡里的心灵原乡。

"湖广填四川"与移民会馆

明末清初，连年战乱的四川，人口凋零，田亩荒废，尤其在成都地区，作为当时各方征战的主要战场，城乡经济更是遭到毁灭性破坏。无力完纳朝廷赋役的川省官守，只好向清廷奏明实情——"蜀民憔悴"，"举目荆榛"，"残民无多……无凭造报"，并提出"整理残疆"，"生聚教养"的建议。

针对四川的这一特殊情况，自清顺治十年（1653年）起，清廷制定并实施了南北各省移民入川垦殖的政策——"准四川荒地，官给牛、种，听兵民开垦"；"凡抛荒土地，无论有主无主，任人尽力开垦，永给为业"；凡开荒者，"暂准五年之后起科"；不拘蜀民、外籍，招民300户者，可"实授本县知县"等。

但此间，四川依然战火频仍。在打败张献忠部之后，因残余抗清势力的疯狂抵抗，清军与之对省会成都及其时川东重镇重庆的反复争夺，持续达20年之久，直到康熙四年（1665年），原设川北保宁（今阆中）的省会才最终迁回成都。而在此后十年，因"三藩之乱"，四川战火再起。从清康熙十三年（1674年），三藩中兵力最强的吴三桂军队攻入四川，到清康熙二十年（1681年）全川平定，七年时间里，本已处在饥荒中的蜀民，再次遭到异常惨烈的战乱蹂躏。吴军"剥肤于疮痍之余"，亘古浩劫，四川枯骨对泣，一片荒凉，甚至"民互为食"，无疑人间地狱。

这使得一度推行的移民策略并无实效。直至康熙二十年（1681年），四川全境肃清，康熙皇帝再推入川垦荒的移民政策，四川生机才逐渐恢复。到清康熙五十一年（1712年），四川便出现了"人民渐增，开垦无遗"的兴盛局面。清雍正元年（1723年），清廷又实行了更为优惠的入川垦殖政策——"奇荒田垦种，六年起科；荒地垦种，十年起科"，这更加吸引了各省移民入川的热情。数以十年的垦荒，使四川再度恢复了沃野千里、人富粮多的风采。虽然自清雍正五年（1727年）起，清廷已经明令停止向川移民，但迄至清乾隆年间，抱着到蜀地发家致富梦想的移民仍然络绎不绝。

清代学者魏源曾对这段历史撰文说，"当明之季世，张贼屠蜀民殆尽，楚次之，而江西少受其害。事定之后，江西人入楚，楚人入蜀"，所以，当时有"江西填湖广，湖广填四川"之说。据史料记载，康熙平乱后发布《招民填川诏》的百年间，先后有623万人迁入四川，其中湖广（明代至清雍正元年（1723年），"湖广"范围为湖南、湖北两省。清雍正元年已降，湖广行省被划分为湖北、湖南两省）移民约346万，占移民总数的55.5%，闽、粤、赣客家移民共247万（广东移民约144万，江西移民约83万，福建移民约20万），占移民总数的39.6%，河南、陕西、山西、山东、安徽、云贵、江浙等其他省份的移民占移民总数的4.9%，所以，清初这场中国历史上最大规模的移民浪潮，便被学界冠以"湖广填四川"。

事实上，中国历史上曾有过两次"湖广填四

川"，第一次发生在元末明初，第二次发生在清代初期，而我们通常说的"湖广填四川"是指的清初这次移民运动，它为四川地域文化作了最近历史最后一次的铺底和构建。

各地移民的到来，使得款叙乡情的共同空间移民会馆在各地应运而生。在成都，因为陕西人最早来此拓荒，所以早在清康熙二年（1663年），成都就有了陕西会馆，之后又陆续建起江南会馆、贵州会馆、湖广会馆、山西会馆等。由于赣南及岭南的客家人，大多是在湖广等籍移民之后落脚四川的，因此客家人的广东会馆、江西会馆、福建会馆都稍迟建立。

在所有迁往四川的移民中，湖广移民为数众多，分布范围广泛，使得四川的湖广会馆一度达172座，占全国总数的78.5%。而江西虽然与四川隔着湖北，但"江西填湖广，湖广填四川"，从数量统计看，四川的江西会馆一度达320座，超过湖广会馆位居第一。除此之外，山陕会馆的关帝庙，一度也遍布城乡。根据清代方志记载，其时的四川无地没有异省会馆，分布密度之高，举国无二。

与此同时，随着四川经济社会的复苏，各路工商人士看到四川百废俱兴的种种商机，也纷纷前来，于是乎，为沟通买卖、保障共同利益的各行业会馆，也顺势在各地建立起来。比如有制盐业的盐神庙，木船运输业的王爷庙、铁匠帮的雷祖庙、屠宰业的桓侯宫、缝纫业的轩辕宫等等，不一而足。这里需要说明的，即便是那些移民会馆，其建造者和管理者，也多为同籍商人的"义举"，只是它们更偏向于乡梓之情的认同，兼顾商务而已。

其时的四川各地，闹市、街巷、乡场，一度会馆林立，它们是四川移民社会形成的标志，也是"天府之国"再度繁荣的标志。

硕果仅存的洛带会馆

"湖广填四川"之所以会引发如此大规模的移民浪潮，与明清时代长江流域与巴蜀地区水运格局的初步形成，有着很大关系。及至明代，湖广地区长江、汉水的江河堤坝建设已基本成形，使得过去水道无常的江河泛滥区，成为中国水运交通的枢纽之地。"湖广通则天下通"，考察各籍移民的入川线路，除甘陕移民通过古蜀道、贵州移民经川黔线入川外，其余大多移民往往都中转于湖广地区，尔后入川的。比如，江西移民主要是从鄱阳湖地区往西进入湖北境内后入川；河南、安徽等地的移民，则沿黄黄官道翻越大别山，再南下沿举河水路进入湖北麻城、黄州后入川；安徽东部移民顺长江、过九江入湖北黄州，之后入川；河南移民顺汉水，过襄阳进入湖北，再入四川；广东、广西、福建的移民也大都先到湖南，之后再入四川。纵观各地会馆之分布，大多地处江河水运畅通之地。因为"不通"则不可能迁移，"不畅"则不能有大规模的迁徙，但过于快捷通畅，则没有必要迁徙，所以当铁路交通出现后，会馆便日趋走向衰落。

当各籍移民来到四川后，出于对地域生活环境的重新认识，势必进行再度迁徙，其中省会成都及近郊是他们的首选，而明末清初，几度战火沦为废墟的成都也正渴望着各方移民前来拓荒生息。

谈及此，一段历史进入我们的视线——明崇祯十六年(1643年)，一度投降明朝廷，又再次反叛的张献忠率六十万大军攻入四川，由于遭到地主武装的蜂起抵抗，张献忠采取"除城尽剿"的政策，大开杀戒，甚至攻克成都之后，纵兵屠城三天；逃离成都时，又屠城"剿洗全城居民"，将偌大一座成都城焚为焦土，火势之猛，"连月不绝"；"官民庐舍，劫火一空"，"人畜同化灰烬"，不仅如此，成都府属州县也大多没有躲过这场劫难。

据清代学者王沄的《蜀游纪略》，在清康熙初年，他到成都看到的场景依然是"人烟久绝，尽成污莱……野外高山累累……城中茅舍寥寥，询其居民，大多秦人矣"，可见陕西人是最早到成都拓荒谋生的，所以陕西会馆早在清康熙二年（1663年）就在成都建立了。之后晋、楚、豫各省也相继而来，并陆续建立会馆。

事实上，先期移民的会馆大多建在成都城内荒野废墟之地，当初并无街道店铺和民居，后来因会馆会期甚多，不少商家小贩便在会馆之旁摆摊设点，渐成街坊，街道也以会馆名。比如陕西街、湖广街、江南馆街、燕鲁公所（直奉会馆）街、贵州馆街等。而其他不少街道，其命名也多与会馆有关，比如成都小天竺街的命名，是因为街北有浙江会馆，成都人称其为小天竺庙，街因此得名；在金玉街，曾有浙江移民修建的浙江会馆，每逢有浙江籍举子考中状元，就会有乡人在会馆中为这位状元挂上一道匾额，浙江会馆匾额之多，让成都文士莫不羡慕，将浙江会馆喻为"金玉满堂"，金玉街因此得名；在棉花街，有湖北黄州移民修建的帝主宫，街巷移民多以到故地黄州贩运棉花为业，街中店铺也多以经营棉纱、棉絮、棉布为主，街因此得名。

除此之外，在总府街，有福建会馆、湖广会馆；在正通顺街，有云南会馆、山西会馆；在古中市街，有山西会馆；在西糠市街，有广东会馆；在布后街，有河南会馆；在卧龙桥街，有川北会馆等。据统计，其时的成都城，大小会馆有近三十座之多。不仅如此，近郊县城，乃至乡场，也是会馆建筑幢幢而立。

不过，随着时间的推移，会馆功能渐失，加之城市建设的需要，不少会馆或拆毁无迹，或零落颓败，或迁往异地，曾经作为移民精神象征和移民历史背影的会馆建筑，正从我们眼皮下消失。然而，成都近郊龙泉山下的洛带却是个奇迹，从上街到下街，广东会馆、江西会馆、湖广会馆，若干年来，它们静卧于古镇的"一街七巷"，就犹如一件件被主人精心收藏的传家古董，当我们陡然发现它们的存在时，那种历经沧桑的历史韵味，远离时空的悠悠之美，着实让人意外，让人惊喜。

洛带会馆的母语光辉

作为中国场镇现存会馆建筑遗存最多的古镇，会馆聚集的洛带，至今仍呈现出过去移民文化风云际会所带来的一种兼容和开放的人文态势和昂扬精神。

成都洛带是中国内陆"最后的客家王国"，据统计，今天生活在这一特殊人文地理单元里的居民，有90%以上的都是客家人后裔。所谓客家，就是"客而家焉"。从晋代始，这支汉民族支系，因战乱、饥荒等原因，一次次离开中原故土，筚路蓝缕，历尽艰辛，行走在寻找新家园的路上，并最终于唐末宋初蛰居在了闽、粤、赣的崇山峻岭中。当北方民族在不断动态融合之时，他们却保持相对静止，外压内聚，寻根报本，逐步演化为汉民族中既保持中原古文化原态风貌，又兼收并蓄南方各少数民族精华、具有独特文化气质的族群。

三百多年前，无数客家人又勇敢地加入到"湖广填四川"的移民浪潮中。由于较早到达的湖广移民占据了土地相对肥沃的平原地带，入川的客家人大多只好在成都以东的丘陵和山地耕种繁衍，经过上百年的发展，学术界称为"东山客家"的这片区域，最终成为以洛带为中心、四川乃至西南地区最大的客家人聚居区。

"宁卖祖宗田，不丢祖宗言"，是客家先祖的遗训。在1946年一个油菜花盛开的春日下午，当著名语言学家董同龢来到"东山客家"一处聚居区时，客家人习以为常的"乡谈"，诸如将"下雨"念着"落水"，"穿衣"念着"着衫"，"太阳"念着"热头"，"一日三餐"念为"食朝、食昼、食夜"等，引起董同龢的浓厚兴趣。通过对3000多个词条及其发音的整理和研究，董同龢惊讶地发现，这种有着独特音韵的方言与客家语代表的广东梅县话也没有太大的区别，更与当时认为早已消失了的中国古代汉语发音有着某种神秘的联系，而这个"古代"，直指中原的唐宋。

对于一个长期动荡迁徙的民系而言，身外之物的"祖宗田"失去可以再得，但母语的遗忘，将会使他们在迁徙的路上成为断了线的风筝一般迷失航向。更为重要的是，铭记了客家人太多族群记忆的客家话，是客家人情感交流的密码，无论走到哪里，那亲切的乡音都会在他们的心底掀起巨大的涟漪，"耳闻乡梓之音，皆大欢喜"，血脉之根也就由此串联开来。所以，学者严奇岩说，"客家方言对于唤醒、凝聚同祖同宗的情感起了不可低估的作用"，"与其说是保存自己的母语，倒不如说是在保护自己的传统文化。"

尤其在强大湖广文化的包围之下，移民会馆的修建对客家人而言，更有一种"不丢祖宗言"不可或缺的需要。在会馆里，客家人"迎麻神，聚嘉会，襄义举，笃乡情"，互诉乡愁，同吐欢欣；尤其是会馆里的乡土神祇，所唤起的那份亲切的原乡情结，不仅使他们恪守祖训，惟礼是尚，更是他们抱聚成团，在陌生土地上再建家园的力量源泉。

其实在洛带，湖广会馆属湘、鄂籍湖广人的移民会馆，只有广东会馆和江西会馆才是广东籍和江西籍移民的客家会馆。至于上场口的川北会馆，那是因城市建设需要，从成都卧龙寺街异地原貌搬迁至此的。据田野调查，在"湖广填四川"的移民浪潮中，最先到达这里的是山陕地区的秦晋商人，过去他们也曾在洛带建有移民会馆秦晋宫，但早已湮灭无痕。秦晋宫的消失，充分说明在当时的洛带移民聚居区，历史变迁中族群势力的流变和文化的消融。不过从某种角度讲，在洛带这个大的客家文化背景下，湖广人也是广义上的客家人（相对于原籍），洛带的湖广会馆也是今天语境下的客家会馆。

在"文革"时期，为了保护会馆，客家人甚至用泥将那些建筑物上带有"封建色彩"的纹饰涂抹一平。因为在客家人心目中，会馆就是他们不能丢掉的"祖宗言"。

今天，经过客家人数百年的辛勤耕耘，当年"阡陌百里，荒无人烟"的荒凉萧条之地，已为硕果累累的"桃红东山"。当带着花香果味的洛带山野之风轻拂而来的时候，与它一起唱和的，是会馆高翘的飞檐下，古老的风铃在低吟浅唱……而如风一样徜徉的，是母语光辉映照下的会馆建筑群落里，客家人独树一帜的风情画卷。

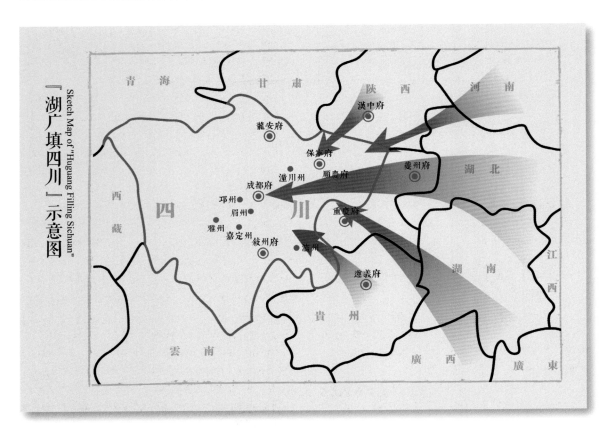

"湖广填四川"示意图

From "Official Mansions" in Han Dynasty
to Luodai Guild Halls

■ SUMMARY

For the non-natives, Luodai is a place where guild halls of migrants gathered. Starting from Shangchangkou, there lie North Sichuan Guild Hall, Guangdong Guild Hall, Jiangxi Guild Hall, and Huguang Guild Hall, respectively. Architectural attractions in different styles and features stand as landmarks fascinatingly and charmingly among the street blocks of the ancient town.

High walls, tranquil yards, overhanging eaves, bucket arches, thick doors, narrow windows, patio, and theatrical stage…for so many years, the guild hall with wide and bright constructional space has witnessed passers-by from far and near and vicissitudes of life; while those exquisite architectures have concealed so many historical and cultural stories.

Looking out from the old-fashioned "nine grid pattern" window, the moon is hanging in the sky tonight…a layer of mystery and early light of glory falls on strangers in a strange land, hometown accent and nostalgia, weather-beaten building community and living things, this is the breath of Hakka migrants and fascinating historical temperament through long time precipitation in the mysterious environment.

Guild halls — hometowns in a foreign land

Throughout the ages, the people traveling across foreign lands would require a place to perch and inhabit. In the historical records, courier hostels for accommodating the people from other lands were set up in some cities as early as Zhou Dynasty. Till Han Dynasty, "official mansions" of each local county were set up in Chang'an. When the bureaucrats from the same hometown came from afar, official mansions full of hometown fellows could offer guests delicious meals in the flavor of hometown. Until Song Dynasty, mansions for serving the guests, merchants, and students from afar could be found everywhere. Meanwhile, these mansions began to divide guestrooms by regions and industries, for instance, there was an "Eagle Store" in south of Panjie Building of Kaifeng City in the Northern Song Dynasty, and it only provided rooms for the customers who traded eagles and falcons. Afterwards, some townsmen associations organized by non-natives with the purpose of seeking common profits along with hometown-fellows in Hangzhou emerged in Lin'an, the capital of Southern Song Dynasty.

Thus, Chinese people always attach great importance to friendships among hometown-fellows and business peers, and taking them as a link for social communication, so as to seek common interests in a foreign land. But properly speaking, they were not like guild halls-official mansions were nothing but equivalent to "Offices in Beijing" nowadays, and those "mansions" seemed more like hotels and hostels nowadays in terms of functions.

Generally, the appearance of guild halls can be traced to the first year of Emperor Yongle's reign of the Ming Dynasty (1402), in which year Zhu Di won the Battle of Jing Nan and ascended the throne, and then became the third generation of emperor of the Ming dynasty. As soon as he ascended the throne, Zhu Di changed the city name from Beiping into Beijing and started a series of constructions for moving the capital city from Nanjing to Beijing, with the purpose of consolidating border defense of the empire against the persistent threat of enemies in the north of China. In fact, the construction of Beijing City took 15 years till the first lunar month of the 19th year of Emperor Yongle's reign (1421), when Zhu Di formally issued imperial edict to move the capital to Beijing. However, Beijing, known as the temporary residence of the emperor, was always the place where Zhu Di was working since the 7th year of the Emperor Yongle's reign of the Ming Dynasty (1403-1424). Therefore, many officials of the Ming Dynasty working in Beijing joined hands with gentries and merchants from their hometowns to raise fund to buy lands in the city and build premises as places for lodging and meeting, so as to go to the court and to have audience with the emperor conveniently.

In the 13th year of the Emperor Yongle's reign of the Ming Dynasty (1415), Zhu Di changed the host city of the triennial metropolitan examination of the imperial examination from Nanjing to Beijing. As a result, tens of thousands people swarmed into Beijing. After the imperial examination, many students who failed to pass the examination preferred to stay in Beijing and wait for the next examination due to the distance from their hometown. Thus, some kind-hearted gentries from the same hometown would invite them to live in the guild halls which they jointly raised funds to build. In the imperial examination period, passing the imperial examination and appearing in the loyal list would be an honor renowned across a place. Therefore, many provinces, state capitals and counties usually competed to build guild halls providing convenient lodging for the students about to take the imperial examinations. But such guild halls, as clearly specified in Record on Reconstruction of Guild Hall in Shexian County, were the "specifically set for the people participating the imperial examination. Traveling merchants have their own apartments, hence they should not live in the guild halls and their goods should not be stalled in here neither, otherwise the intention of chivalrous deeds would fail". Therefore, the initial guild halls were called "trial guild hall" or "scholar guild hall." However, there were subtle differences between them. Generally, "guild halls within the urban area usually accommodate gentries are the hosts; while guild halls outside the area usually accommodate candidates for the imperial examination."

At the same time, some commercial guild halls serving the hometown fellows and the same industry were also built in water transport hubs and bustling commercial ports in each region. For guild halls in the influx of traveling merchants, they were more like a kind of protest and resistance against the exclusion of guild halls only for intellectuals than an imitation by merchants. In the traditional Chinese society, people were divided into four kinds—"intellectuals, farmers, workers and merchants". The merchants were ranked last. It can be seen that for the merchants, the guild hall is an approach to seek social status recognition for them by means of "hometown friendship" and "gather to prosper". Therefore, at the beginning of the operation of the commercial guild hall in each region, strict "regulations", "constitutions", "customary rules", etc., were formulated. Their purposes, of course, were "gathering merchants together, flocking the hometown fellows, openly discussing business, helping colleagues" as well as making benefits and enhancing relationship. The "benefits" here referred to commercial profits, while the "friendship" stood for the essence of Confucianism. Thus, the guild halls were not only the social container combining the interests of the industry and commerce as well as Confucian moral concepts. To some extent, the self-management close to the ideology of the upper society not only conforms to the social management needs, but also the basis for stability of the social order at that time. Therefore, in the narrow sense, the guild hall was an architectural form; while in a broader sense, the guild hall was a social organization.

It can clearly be seen that the guild halls were built based on the geopolitics as the link, strangers in a strange land while hometown accent and nostalgia, so-called "Cherish friendship among relatives and neighbors, narrate homeland friendship, just like hometown fellows though in different places". Therefore, the guild hall is the soul of the original village for the refugees in a foreign land.

"Huguang Filling Sichuan" and Migrant Guild Halls

From the end of Ming Dynasty and the beginning of Qing Dynasty, the population was decreasing and the crop land was dilapidated, especially in Chengdu. As a main battlefield of several parties at that time, Chengdu's urban and rural economy was devastatingly damaged. Officials of Sichuan province who cannot achieve the taxes specified by the court could only reported the actual situation—"Sichuan people are thin and pallid", "desolation can be seen almost everywhere", "the population is too few to afford the tax", etc. And the suggestions such as "gathering people to rehabilitate" were proposed.

For the special case in Sichuan, since the tenth year of the period under the reign of Emperor Shunzhi (1653) of the Qing Dynasty, the Qing government formulated and implemented policies of carrying out cultivation by migrants from other provinces to Sichuan—"waste land are cultivated by soldiers and farmers, while the cattle and seeds are provided by the government". "Where there are abandoned or wasted lands, with or without owner, people are allowed to do their best to cultivate and claim ownership upon the land permanently". A person might even be assigned as the magistrate of a county if he can attract 300 households from Sichuan or other provinces to settle down.

But during that period, the war was still raging in Sichuan. After the defeat of the Army of Zhang Xianzhong, due to heavy resistance of the residual anti-Qing insurgency, the conflicts over the provincial capitals, Chengdu and Chongqing, had lasted for 20 years. Until the fourth year under the reign of Emperor Kangxi of the Qing Dynasty (1665), the original provincial capital, which was set in Baoning in north Sichuan (Langzhong today) was finally moved back to Chengdu. In the next decade, because of the "Revolt of the Three Feudatories", war broke out in Sichuan again. From the thirteenth year under the reign of Emperor Kangxi of the Qing Dynasty (1674), when the strongest force among the three feudatories—Wu Sangui's troop invaded Sichuan, to the twentieth year under the reign of Emperor Kangxi of the Qing Dynasty (1681) when the whole Sichuan was suppressed. During the seven years, Sichuan people who were already starving were once more ravaged by the tragic war. With Wu Sangui's troop "peeling the skin from the wounds", there were skeletons nearly everywhere; and people even "began to eat each other". No doubt that Sichuan was a hell on earth.

That made the implementation of migration policy ineffective. By the tenth year under the reign of Emperor Kangxi of the Qing Dynasty (1681), the whole territory of Sichuan was eliminated. Emperor Kangxi re-executed migration policy to Sichuan, and vitality in Sichuan gradually restored. By the 51st year under the reign of Emperor Kangxi of the Qing Dynasty, prosperous situation of "population increasing, cultivation expanding" occurred in Sichuan. In the first year under the reign of Emperor Yongzheng of the Qing Dynasty (1723), the Qing government implemented a more favorable migration policy in Sichuan—"abandoned fields shall be cultivated with taxation postponed to the 6th year of cultivation; waste land shall be cultivated with taxation postponed to the 10th year", which attracted more migrants from other provinces. Decades of reclamation had led to fertile lands, well-off people and abundant grains again in Sichuan. Although in the 5th year under the reign of Emperor Yongzheng of the Qing Dynasty (1727), the Qing government explicitly proclaimed to stop migration to Sichuan, the migration wave never stopped untill the period under the reign of Emperor Qianlong of the Qing Dynasty, with countless migrants coming to Sichuan, dreaming of getting rich.

Scholar Wei Yuan in the Qing Dynasty wrote an article on this period of history, saying "in the late Ming Dynasty, Zhang Xianzhong, a leader of peasant uprisings, massacred many people, among which the victims the Shu people were the majority, next were the Chu people, and the Jiangxi people were the least. Till the early Qing Dynasty, the Jiangxi people began to move into the state of Chu and the Chu people to

the state of Shu". So at that time there was a saying of "Jiangxi filling Huguang, Huguang filling Sichuan." According to historical records, 6.23 million people migrated to Sichuan 100 years after the Imperial Decree to Fill Sichuan was issued by Emperor Kangxi after his triumph over a riot. Huguang (from Ming Dynasty to the first year (1723) of Emperor Yongzheng's reign in Qing Dynasty) includes Hunan and Hubei Provinces. In the first year of Emperor Yongzheng's reign in Qing Dynasty, insurgents surrendered and Huguang was separated into two provinces, namely, Hunan and Hubei. The number of migrants from there was about 3.46 million, taking up 55.5% of all the migrants. The Hakka migrants from Fujian, Guangdong and Jiangxi were 2.47 million (about 1.44 million from Guangdong, 830,000 from Jiangxi and 200,000 from Fujian), occupying 39.6% of all migrants. Migrants from other provinces like Henan, Shaanxi, Shanxi, Shandong, Anhui, Yunnan, Guizhou, Jiangsu and Zhejiang accounted for 4.9% of the total number of migrants. Therefore, this largest wave of migrants in the beginning of Qing Dynasty in Chinese history has been referred to as "Huguang Filling Sichuan" by the academia.

As a matter of fact, there were two waves of "Huguang Filling Sichuan" in Chinese history. The first one came at the turning point of Yuan Dynasty and Ming Dynasty. The second one took place in the early days of Qing Dynasty. "Huguang Filling Sichuan", generally speaking, refers to the migration wave in early Qing Dynasty, which paved the way for the origin of Sichuan's regional culture for the first time in recent history.

It is because of the arrival of migrants from different places, different kinds of migration guild halls (common spaces for people from the same places) began to appear on the scene. Shaanxi people were the earliest pioneers came to Chengdu. As a result, the Shaanxi Guild Hall appeared in Chengdu in the second year under the reign of emperor Kangxi (1663). Afterwards, Jiangnan Guild Hall, Guizhou Guild Hall, Huguang Guild Hall, Shanxi Guild Hall, etc. were built gradually. Since the Hakka people in Gannan (the south of Jiangxi) and Lingnan (the area south of the Five Ridges) mostly arrived in Sichuan after Huguang and other migration waves, so their Guangdong Guild Hall, Jiangxi Guild Hall and Fujian Guild Hall were built later.

Of all the migrants who moved into Sichuan, the number of Huguang migration was the most and they distributed in a wide range. Hence the number of Huguang Guild Halls in Sichuan had once reached up to 172, accounting for 78.5% of the national total amount. Although there was Hubei between Jiangxi and Sichuan, as the saying "Jiangxi filling Huguang, Huguang filling Sichuan" and according to the quantity statistics, Jiangxi Guild Halls in Sichuan were still once up to 320, even more than Huguang Guild Halls and was ranked first. Besides, the Guan Yu Temples of Shanxi and Shaanxi Guild Halls were also built everywhere throughout urban and rural areas. According to local records of Qing Dynasty, at that time migration guild halls were everywhere in Sichuan, the density of distribution was high. What's more, such situation was unique in the whole country.

At the same time, with the resurgence of Sichuan's economy and society, many industrial and commercial people realized a variety of business opportunities in the process of change in Sichuan, so they also came. As a result, for the communication of buying and selling, safeguarding the common interests of industry, business guild halls were also set up. Such as salt god temple for salt producing, Wangye temple for wooden ship transportation, Leizu temple for blacksmith party, Huanhou palace for slaughtering, Xuanyuan palace for sewing and so on. Here need to note, the builders and managers of those migration guild halls, mostly are also businessmen from the same place. Their behaviors of building and managing were treated as acts of kindness. But compared with business guild halls, the migration guild halls are more prefer to fellow villagers' identities than to business.

At that time guild halls were all over villages, towns and streets in Sichuan. They were markers of establishment of Sichuan migrant society and the sign of prosperity in the "land of abundance".

The Only Remaining Luodai Hall

The reason that "Huguang filling Sichuan" led to such a large wave of migrants, had a great relationship with the initial form of the waterway pattern of Yangtze River and Bashu regions in the Ming and Qing Dynasties. Until the Ming Dynasty, in Huguang regions, the dam construction of Yangtze River and Han River had been basically formed, which transformed the past river flood zone with impermanent waterways into a hub of China's water-land transportation. "Only Huguang lines open can the whole country be connected". According those "lines into Sichuan", migrants with different native places, expect those migrants who from Gansu and Shaanxi provinces by the ancient Sichuan Road, as well as those migrants from Guizhou province via the Sichuan-Guizhou line, most of the remaining migrants often transited into Huguang areas, and then entered Sichuan province. For example, migrants from Jiangxi province mainly started from the Poyang Lake region, going west into the territory of Hubei then enter Sichuan; migrants from Henan, Anhui and other provinces climbed the Dabie Mountain along Huanghuang Road, then went south along the Juhe River into Macheng and Huangzhou in Hubei province, then entered Sichuan; migrants from east Anhui went along the Yangtze River, went over Jiujiang River and went into Hubei province's Huangzhou, then entered Sichuan; migrants from Henan province went along the Hanshui River, went through Xiangyang and reached Hubei, then entered Sichuan; migrants from Guangdong, Guangxi, Fujian provinces were mostly first going to Hunan, then entered Sichuan. Looking around the distribution of halls, most are located in the unblocked place of water transportation. Because "blocked" meant impossible migration, and "obstructed" would cut off a large-scale migration, but if the transportation was too convenient, then the migration would be unnecessary, so when rail traffic appeared, the halls became increasingly toward decline.

After entering Sichuan, out of a new understanding of the local living environment, migrants with different native places were bound to re-migrate. And the provincial capital Chengdu with its suburbs was their first choice. While in Ming and Qing Dynasties, Chengdu, suffering from damages during several wars, was still also expecting the migrants from different provinces to cultivate lands and settle down.

Turning to this, a piece of history enters our line of sight-in the 16th year under the reign of Emperor Chong Zhen if Ming Dynasty, Zhang Xianzhong, who once surrendered to the Ming court and then rebelled, led an army of 600,000 to invade Sichuan. Because of the armed landlords' resistance, Zhang Xianzhong adopted a policy of "destroy the city and raid all". Even after occupying Chengdu, the army massacred for three days; when they fled from Chengdu, they had not only "raided the whole city's residents", but also burned the huge Chengdu city into scorched earth. The fierce fire burned for months without break; "the official and civilian farmhouses had been robbed of everything"; "human beings and livestock were reduced to ashes". Moreover, counties belong to Chengdu mostly did not escape this catastrophe as well.

According to A Diary of Traveling in Shu by scholar Wang Yun in Qing Dynasty , who at the beginning of years under reign of Emperor Kangxi of Qing Dynasty went to Chengdu and saw the scene of "the population was sparsely distributed, all turned into wastelands...there are only few huts in town. I went to inquire the inhabitants, most of whom are Qin people", which showed that Shaanxi people were the earliest migrants to Chengdu to cultivate lands and settle down. Therefore, Shaanxi Guild Hall was built up as early as in the second year under the reign of Emperor Kangxi of Qing Dynasty (1663) in Chengdu. Thereafter, Jin, Chu, Yu and other provinces followed to Chengdu, and established their guild halls.

In fact, the early guild halls were mostly built upon wastelands in Chengdu downtown where there were no streets, shops and dwellings then. But later, for the guild

halls' frequent meetings, many merchants and vendors set up stalls beside guild halls which made them become neighbors, so streets were named after the halls. For example, Shaanxi Street, Huguang Street, Jiangnanguan Street, Yanlugongsuo (Zhifeng Guild Hall) Street, Guizhouguan Street, etc. And the nominations of many other streets were also more or less relevant with guild halls, such as Chengdu's Xiaotianzhu Street. Because there was a Zhejiang Hall on the north street which Chengdu people called Xiaotianzhu Temple, the street got its name. On Jinyu Street the Zhejiang Hall was built up by Zhejiang migrants. Whenever there was a Zhejiang candidate ranked first in the imperial examination, the local people would put up a plaque for the candidate in the guild hall. The number of plaques in Zhejiang Guild Hall was so huge, which made Chengdu intellectuals admire so much that they hailed Zhejiang Guild Hall as "a hall filled with gold and jade" hence there is the Jinyu Street. On the Cotton Street, there was an emperor palace built by migrants from Huangzhou, Hubei. Many Huangzhou merchants transported cotton for sale and shops were mostly trading veil, cotton batting and cotton cloth so the street got its name.

In addition, on Zongfu Street, there were Fujian Guild Hall, Huguang Guild Hall; on Zhengtongshun Street, there were Yunnan Guild Hall and Shanxi Guild Hall; on Guzhongshi Street, there was Shanxi Hall; on Xikangshi Street, there was Guangdong Guild Hall; on Buhou Street, there was Henan Guild Hall; on Guwolong Bridge Street, there were North Sichuan Guild Hall and so on. According to statistics, the city of Chengdu at that time had nearly thirty large and small guild halls. Moreover, in suburban counties and even towns, blocks of guild hall buildings were also built up.

However, as time went by, with the gradual loss of guild hall functions and the needs of urban construction, many halls were either demolished without a trace, or scattered decadent, or moved to different places. The guild hall buildings once being symbols of the migrant spirit and migration history began to disappear. However, Luodai, located at the foot of Longquan Mountain in the suburb of Chengdu, was a miracle. From the upper street to the lower street, lied Guangdong Guild Hall, Jiangxi Guild Hall and Huguang Guild Hall. For a number of years, they repose in the ancient town's "one street and seven lanes", like pieces of heirloom antiques collected carefully by the owner. When we suddenly discovered their existence, the charm of historical vicissitudes and the distant beauty of hiding away from space and time, indeed surprised and pleased us a lot.

Glory of Luodai Guild Hall's Mother Tongue

As an ancient town with the largest number of existing hall architecture in China's rural farms, Luodai with gathered guild halls has also shown compatible and open humanistic trend as well as high-spirited passion brought about by the migrant culture.

Chengdu's Luodai is "the last Hakka Kingdom" in China's inland. According to statistics, today the residents live in this exceptional human geographic unit, among whom more than 90% are of Hakka descents. The so-called Hakka is "living in a foreign land as in the hometown." From the beginning of Jin Dynasty, this branch of the Han nationality, because of the war, famine and other factors, left their homeland in Central Plains, arduously walking on the road to look for a new home. Finally, in the end of Tang Dynasty and the beginning of Song Dynasty, they settled down in seclusion in the high mountains in Min, Yue and Gan areas. When the northern nations were in constant dynamic fusion, they remained relatively stationary. With external pressure and internal cohesion, they sought roots, and gradually evolved into an ethnic group with unique cultural qualities in Han nationality, which not only maintains the original state of ancient culture style in the Central Plains, but also all-embraces the essence of eclectic Southern China ethnic groups.

Three hundred years ago, many Hakka again bravely joined the wave of migrants, "Huguang filling Sichuan". As Huguang migrants who arrived earlier occupied the relatively fertile plain lands, most of those Hakka had to farm and reproduce upon hills or mountains in the east of Chengdu. After a century of development, this region which called by academia as "Dongshan Hakka", ultimately become the largest Hakka community which took Luodai as the center in Sichuan and southwest region.

This sentence "I could sell forefathers' farms, not would never forget their words" was one of Hakka ancestors' teachings. In 1946, a spring afternoon with blooming canola flower, when the famous linguist Dong Tonghe came to a ghetto of "Dongshan Hakka", the "hometown talk", such as saying "xia yu (raining)" as "luo shui (落水)", "chuan yi (dressing)" as "zhuo shan (着衫)", "tai yang (the sun)" as "re tou (热头)" as well as the three meals a day as "shizhao (食早)", "shizhou (食中)" and "shiye (食夜)", etc., which Hakka people took for granted triggered his strong interest. Through the collation and research on more than 3,000 entries and pronunciation, Dong Tonghe to his surprise found that this dialect with a unique phonology had not much difference from the Guangdong Meixian dialect which was a representative of Hakka dialect, and also had some mysterious connection with the ancient Chinese pronunciation of Chinese, which at that time had disappeared. And this "ancient", directed at the Tang and Song dynasties in the Central Plains.

In fact, for an migrant group which was for a long-term in turbulence, the mere worldly possession of "ancestry farm" could be regained after being lost, but the forgotten mother tongue would make them lost along the road of migration. More importantly, the Hakka dialect which contained many Hakka ethnic memories was an emotional communication password for Hakka. Wherever they went, that kind of cordial accent will be in their hearts setting off huge ripples. "Hearing hometown voice, all are in happiness" that the root of blood from this connected in series. Therefore, scholar Yan Qiyan said, "The Hakka dialect in waking and cohering the same ancestral emotions played an underestimated role", "it was more than saving their mother tongue, than protecting their traditional culture."

In particular, under the strong Huguang culture's siege, the construction of migrant guild hall, for the Hakka people, has an essential need to "do not lose forefathers' words". In guild halls, the Hakka people would "honor the Ma God, have gathering, cooperate for chivalrous deed, consolidate home-town fellowship" as well as exchange nostalgia and joy. Especially for the cordial homeland complex evoked by the local gods in guild halls that not only enabled them to abide by the teachings of forefathers, but also to hold together as a group, so as to build a better home in a strange land.

In fact, in Luodai, Huguang Guild Hall belonged to migrants from Hunan and Hubei provinces, only Guangdong and Jiangxi guild halls were the Hakka guild halls that belong to migrants from Guangdong and Jiangxi provinces. As for Shangchangkou's North Sichuan Hall, it was directly moved from Chengdu's Wolong Temple Street to here due to urban construction needs. Accoding to field surveys, in the wave of migrants" Hug uang filling Sichuan", the first to arrive here were Qin and Jin merchants from Shanxi and Shaanxi regions, in the past, they built a Qin Jin Palace in Luodai, but the guild hall had disappeared without traces. The disappearance of Qin Jin Palace sufficiently demonstrated the rheology and decline of a culture and the identity of another strong culture in Luodai migrant community. But in some ways, in the large Hakka cultural background of Luodai, Huguang people could also, in a broad sense, be deemed as Hakka people (as their own origin). Huguang Guild Halls in Luodai were also Hakka Guild Halls in today's context.

During the "Cultural Revolution" period, in order to protect guild halls, the Hakka people used mud to cover those ornamentations with a "feudalistic color" on buildings. Because in the eyes of Hakka people, halls are the "forefathers' words" that they cannot lose.

Today, after centuries of Hakka's hard work, the bleak and desolate land with "hundreds of miles of lands deserted and sparsely populated" had been the fruitful and abundant land. When the mountain wind of Luodai with floral fruity flavor blows, what singing along with it is the old wind chime under the raised brackets...while what roaming like wind is the Hakka's unique picture of guild halls cluster under the mother tongue's brilliant resplendence.

全国重点文物保护单位洛带会馆
Important Heritage Site under State Protection Architectural Complex of Luodai Guild Halls

洛带会馆位于四川省成都市龙泉驿区洛带镇，包括湖广会馆、江西会馆、广东会馆和川北会馆。作为当年各籍移民集会议事和祭祀的重要场所，洛带会馆建筑群是中国移民文化和聚落景观的一个代表。

洛带四大会馆布局考究，外廓宏伟壮观，内部构件细腻精巧。洛带会馆极高的建筑艺术价值，为中国会馆建筑的研究，提供了珍贵的样本资料。2006年，洛带会馆被列为第六批全国重点文物保护单位。

Luodai Guild Halls located in Ancient Luodai Town, Longquanyi District, Chengdu City, Sichuan Province, include Huguang Guild Hall, Jiangxi Guild Hall, Guangdong Guild Hall, and North Sichuan Guild Hall. As an important venue for migrants to gather, comment, and worship, architectures of Luodai Guild Halls are one of the representatives of Chinese migrant culture and settlement landscape.

Much attention was paid to the layout of the 4 guild halls in Luodai and they have magnificent external contours and exquisite interiors. Architectural complex of Luodai Guild Halls is of extremely high agricultural value and function as a precious sample for the research on Chinese guild hall architectures. In 2006, Luodai Guild Halls were listed among the 6th group of national key cultural relic protection sites.

　　三百多年前，无数客家人勇敢地加入到"湖广填四川"的移民浪潮中。从中原到岭南，再岭南而四川，客家人从千年之前的历史深处走来，从千里之外的故土家园走来，筚路蓝缕，历尽艰辛，最终像一粒粒饱满的种子，扎根在了蜀地洛带这片茂密的土地上，繁衍生息，开花结果。他们是洛带会馆的营建者和守护者。图为洛带广东会馆大殿临街后墙上浮雕的"客家填川形胜图"。

　　Three hundred years ago, a lot of Hakka people showed their braveness to flood into the wave of migration "Huguang Filling Sichuan". From Central Plain to south of the Five Ridges to Sichuan, the Hakka people, who have passed through thousands of years in Chinese history, left their homes thousands miles away. After enduring great hardships and going through much suffering, just like plump seeds rooted in the vibrant land, they settled down in Luodai in Sichuan to fill up the ancient town and become the builders and protectors of guild halls in Luodai. The picture shows a Migration Map of Hakka People to Sichuan, which is embossed on the rear wall of Guangdong Guild Hall.

　　随着清初各地移民的到来，地处川渝古商道上的洛带再度繁华起来。于是，那些出于联系乡谊，祭祀议事，显耀门庭的移民会馆便在商贾兴隆的古镇次第建成。

　　With the migrant wave during early Qing Dynasty, Luodai, located at the ancient trade route between Sichuan and Chongqing, once again began to prosper. As a result, more and more guild halls were built in the town for people to get in touch with home-town fellows; to convene ceremonies and conferences; and to show off the owner's wealth.

风貌有别的洛带四大会馆，风情洋洋地耸立在洛带古镇的各个街段，它们与簇拥的民居檐壁相接，互为依托，完美地构建出古镇洛带独特的移民记忆和聚落文化景观——会馆宏大的建筑，是移民入川先祖精心移植的原乡记忆；会馆之外，则是热闹的街景，以及为每一户人家挡风遮雨的青瓦民居。

The 4 guild halls with different styles stand on different streets in Luodai. These guild halls are surrounded by dwellings and they are complemented by each other; a perfect landscape of migrant memory and settlement culture that was only seen in Luodai was formed. The large architectures show those pioneering migrants' memory of their hometowns; beside those guild halls, there are lively streets and people's grey-tiled dwellings that keep out wind and rain in Luodai.

第一章
洛带古镇

府之邑曰灵泉，而邑之镇曰洛带 ——（北宋）张溥《灵泉县瑞应院祈雨记》

洛带古镇与洛带会馆

第一章

今天所能查阅到有关"洛带"的最早记载，是唐末五代道教学者杜光庭所撰的《神仙感遇记》——"成都洛带人牟羽宾家贫，鬻力于市"。而洛带最早以建制之镇出现在古籍中的，则是北宋名士苏恽撰于北宋皇佑年间（1049年——1054年）的《灵泉县圣母堂记》中的一段文字——"灵泉邑北，直向驰道……地聚洛带镇市"。之后名士张溥撰于北宋熙宁七年（1074年）的《灵泉县瑞应院祈雨记》中，也写到"府之邑曰灵泉，而邑之镇曰洛带"。

至于"洛带"地名的来历，明代当地文人柳溪的《瑞应寺（今洛带燃灯寺）八景诗》，曾列"寺面孔明落带镇"，依清《简州志》的解释，为"相传武侯落带于此，因名"。而洛带广东会馆有（楹）联曰"恩流洛水"，这里的"洛水"，是指古镇西边绕镇而过的一条小溪，那么洛带得名的本初，是因为这条给当地百姓带来千年福祉的盈盈溪流吗？这就不得而知了。

不过有一点可以明确的，洛带历史的深厚底蕴，实在得源于那条起于汉，兴于唐的川渝古商道。作为川渝古商道上的一个节点，相传洛带早在汉时便聚村成街，有吴泰集、万泰驿、万井街、安镇等多种称谓。三国时蜀汉丞相诸葛亮兴市，洛带以"万福街"名。

而至明末清初，因为连年的战乱和天灾，洛带与蜀地其他地方一样，是一片的凋敝和荒凉。及至清代中期，通过"湖广填四川"移民上百年的艰辛经营，古老的川渝商道上，逐渐流行起了一句谚语——"填不满的牛市口，扯不空的甑子场"，这里所说的甑子场，就是洛带镇，从中我们可以想见，其时的洛带作为商品聚散地的繁华。值得一提的是，古镇当时还以出产客家人用于蒸饭的竹编生活用具甑子，一度以"甑子场"之名享誉一方。

于是，那些出于祭祀议事、招徕同乡、显耀门庭的移民会馆，便在洛带次第建成。根据史料，清乾隆八年（1743年）湖广会馆建成；清乾隆十一年（1746年）广东会馆建成；清乾隆十八年（1753年），江西会馆建成。随着会馆的落成，洛带场镇风貌大为改观，商贾兴隆，名扬周遭，一度别称"甑子场"。

会馆之内供奉有各籍移民信奉的乡土神祇；会馆之外，则是为每一户人家挡风避雨、前铺后宅的青瓦民居。会馆与民居檐壁相接，互为依托，完美地构建出古镇洛带独特的移民文化气质和聚落文化景观。

CHAPTER 1
Ancient Luodai Town
and Luodai Guild Halls

"Coming from a poor family in Chengdu Luodai, Mou Yubin makes a living through hard labor."–This is the earliest reference of "Luodai" that can be found today, quoted from God's Patronage by Du Guangting, an educator of Taoism of Five Dynasties in late Tang Dynasty. Luodai was firstly mentioned as a town in historical document Record of Goddess Palace of Lingquan County, which was written by the celebrity Song Yun during the reign of Emperor Huangyou of Northern Song Dynasty (1049–1054), saying "In the north of Lingquan County, there is a convenient post road leading to Luodai Town". After that, the celebrity Zhang Pu also wrote a sentence "The county of the region is called Lingquan, while the town of the county is called Luodai" in the book Record of Praying for Rain in Ruiying Courtyard of Lingquan County, which was complied in the 7th year during the reign of Emperor Xining of Northern Song Dynasty.

As for the origin of the name 'Luodai', Eight-Scene Poetry of Ruiying Temple (today's Randeng Temple) written by local literati Liu Xi in Ming Dynasty once read "The temple faces a town where Kong Ming dropped his waistband"; according to Jianzhou Chronicles in Qing Dynasty, "the town was named for Wuhou marquis (Kong Ming) dropping his waistband there". Moreover, Guangdong Guild Hall in Luodai has a couplet, saying "Grace to Luoshui River", and Luoshui River refers to a brook in the west flowing around the town. Therefore, did the name itself "Luodai" originate from the flowing river that brings local people prosperity in the past thousands of years? It is still an unanswerable question.

But one thing is clear: the profound historical connotation of Luodai was derived from the ancient Sichuan-Chongqing trade route, which emerged in Han Dynasty and prospered in Tang Dynasty. As a node on the ancient trade route, Luodai, according to legend, was developed into a town by merger of villages and was known as Wutaiji, Wantaiyi, Wanjing Street, Anzhen Town, etc. During the Three Kingdoms period, Prime Minister Zhuge Liang of Shu Kingdom implemented revitalization policy and named Luodai "Wanfu Street".

In late Ming and early Qing Dynasties, Luodai, like other places in Sichuan, suffered from years of wars and disasters, showing a scene of waste and destitution. In middle Qing Dynasty, through over one hundred years of efforts made by the migrants of "Huguang Filling Sichuan", a saying "Zhenzichang is always crowded, just like Niushikou can never be filled" began to spread along the ancient Sichuan-Chongqing trade route. The abovementioned Zhenzichang refers to Luodai Town. It is not hard to picture how a prosperous place like Luodai was as a distribution hub of commodities. It is noteworthy that the town was renowned as "Zengzichang" for it produced Zengzi, a bamboo-woven instrument used by Hakka people to cook rice.

As a result, migrant guild halls emerged one after another in the town of prosperous business. Some were established for rituals and official business, some for solicitation of townees, and some for demonstrating the glorious past of their ancestors. According to historical records, Huguang Guild Hall was established in the 8th year (1743) of Emperor Qianlong in Qing Dynasty; Guangdong Guild Hall was established in the 11th year; and Jiangxi Guild Hall in the 18th year (1753). As guild halls rose, Luodai Town's atmosphere greatly improved and was called "Zhenzichang" for prosperous business and wide reputation.

Inside the guild halls, local gods worshiped by migrants of different ethnicities are enshrined; on the outside are grey-tile dwellings of Hakka families lining up like ranges of hills. Connecting with one another, guild halls and dwellings perfectly depict a unique quality of migration and cultural cluster phenomenon of Luodai ancient town.

随着移民及移民文化的流变，客家文化成为洛带移民社会的主流。在洛带场镇的两万居民中，有90%以上的都属客家人后裔，那些来自原乡的乡音俚俗，至今还在客家人子孙今天的生活中，秉承原样且虔诚地代代沿袭。这一点，会馆功不可没——在会馆乡谊感染和礼俗规范下，客家人在一种独立而自由的社会空间里，抱聚成团，共克时艰，共享繁荣，不自觉地就留存了记忆，保持了经验，维持着客家文化的昌明发达。

During the change of migrants and their culture, Hakka culture became the mainstream of Luodai community. Among the 20,000 residents of Luodai Town, more than 90% of them are descendants of Hakka people, speaking their home tongues and carrying their customs generation after generation. The guild halls, which contributed a great share in preserving the Hakka culture, helped the Hakka people gather together in an independent and free social space to live through thick and thin. By such, they unintentionally kept their tradition, maintaining the thriving and flourishing of Hakka culture.

洛带古镇是中国乡场现存会馆最多的镇，从上场口到下街，依次是川北会馆、广东会馆、江西会馆和湖广会馆。虽然历尽沧桑，但岁月却赋予它们浓郁的历史韵味和悠远之美。

Luodai Ancient Town has the most existing guild halls among villages and towns in China. From Shangchangkou to Xiajie Street, there lies North Sichuan Guild Hall, Guangdong Guild Hall, Jiangxi Guild Hall and Huguang Guild Hall respectively. Though having been through so many decades, they still maintain a profound beauty

第二章
湖广会馆

龙船头上二十四把桨，撑起船来到湖广 ——成都童谣

湖广移民与湖广会馆

第二章

据任乃强先生考证，因为连年战乱、饥荒等因素，"至清顺治七年（1651年），蜀人大体已尽"，为使荒芜的蜀地再现生机，清政府采取了宽松的移民政策，从而出现了"始于顺治末年、盛于康雍乾、止于嘉庆年间的移民浪潮"。

在这次被称为"湖广填四川"的中国历史上最大规模的移民浪潮中，有百万之众的荆楚两湖的湖广移民涌入四川，居其他省籍移民数量之首。湖广会馆随之遍布四川各地。据统计，全国的湖广会馆总计219所，其中四川就占了172所，占总数的78.5%。

湖广会馆是湖广移民的省级会馆，除此之外，还有地方级的会馆，比如湖北馆、黄州馆、武昌馆、郴州馆、长沙庙、衡州会馆、常澧会馆、三楚宫、靖天宫、帝主宫、威远宫等。因为成都是四川的政治经济中心，也是湖广移民的主要迁入区，所以成都一地的湖广移民会馆一度为数众多，比如有洛带的湖广会馆，新都的郴州馆，新津的黄州馆、禹帝宫，邛崃的三楚公所，金堂的靖天宫、帝主宫等。

洛带湖广会馆地处洛带古镇中街，是成都地区至今保存最为完好的湖广会馆之一。会馆建成于清乾隆八年（1743年），坐北朝南，由门楼、戏台、中后殿、前院空坝及两边厢房、两边庑廊围合的四合天井，及东西套院、后院组成，建筑面积达2480平方米。

洛带湖广会馆曾供奉禹王宝相，同时由于关羽生前曾驻军湖北江陵、湖南长沙，最后战死麦城，湖广移民为使乡党增添光彩，也曾塑关羽之像于会馆殿堂。除此之外，还有观音、鲁班、阿弥陀佛等。从单一的乡土神祀奉，到以乡土神为主的多神兼祀，表现出其时生活在洛带一地的湖广移民，在包容了同乡多样祈求的同时，也随着在蜀地的发展，呈现出极强的实用功利。

值得一提的是，洛带湖广会馆后院庑廊，曾是移民寄存棺柩的地方。对于流寓在外的人们来说，偶然的遭遇，会让他们的富贵贫贱变得不可预料，所以在过去，不少移民会馆都设有专为贫困移民寄厝棺柩的地方，对那些无力安葬亲人的移民，还辟有义冢。而这样的慈善之举，无疑大大强化了会馆在乡人心中的地位，从中而见会馆组织对社会各阶层的包容，以及对中国传统伦理道德维系的自觉，并在此基础上参与社会管理的作用所在。

CHAPTER 2
Migrants from Huguang
and Huguang Guild Hall

According to textual researches of Ren Naiqiang, due to the factors such as continuous wars and starvations, up to the 7th year of Shunzhi era (1651), people almost died out in Sichuan. In order to bring vigor back to the deserted Sichuan, the Qing Government implemented migrant policies more flexibly, thus appearing "the migrant wave which started at the end of Shunzhi era, prospered in Kangxi, Yongzheng and Qianlong eras, and ended in Jiaqing era".

Called as "Huguang filling Sichuan", this is the largest migrant wave in the Chinese history, in which millions of people from Hunan and Hubei flowed into Sichuan, ranking the first among migrants from other provinces. Guild halls of Hunan and Hubei were distributed in every place of Sichuan. According to statistics, there were 219 Huguang Guild Hall in China, among which 172 were in Sichuan, taking up 78.5% of the total number.

Huguang Guild Hall is the provincial-level guild hall for migrants from Hunan and Hubei. In addition to it, there are many other local guild halls, including Hubei Guild Hall, Huangzhou Guild Hall, Wuchang Guild Hall, Chenzhou Guild Hall, Changsha Temple, Hengzhou Guild Hall, Changfeng Guild Hall, Sanchu Palace, Jingtian Palace, Dizhu Palace, Weiyuan Palace, etc. As Chengdu is the political and economic center of Sichuan, it is also the main migration area of migrants from Hunan and Hubei. Therefore, there is a large number of Hunan and Hubei migrant guild halls in Chengdu, for example, Huguang Guild Hall in Luodai, Chenzhou Guild Hall in Xindu, Huangzhou Guild Hall in Xinjin, Palace of Yu the Great, Sanchu Guild Hall in Qionglai, Jingtian Palace and Dizhu Palace in Jintang, etc.

Located in the Zhongjie Street of Ancient Luodai Town, the Huguang Guild Hall in Luodai is one of the best preserved Huguang guild halls in Chengdu. Completed in the 8th year of Qianlong era (1743), it is seated in the north and facing the south. It consists of decorated archway, opera stage, middle and rear halls, empty dam of the front yard, the courtyard formed by wing-rooms and wing-corridors, suit yards and rear yards in the west and east. It covers a construction area of 2,480 m^2.

Once, the Huguang Guild Hall in Luodai consecrated the portrait of Yu the Great, but as Guan Yu had stationed his troops in Jiangling of Hubei and Changsha of Hunan, and finally sacrificed himself in Maicheng, migrants from Hunan and Hubei also placed the portrait of Guan Yu in the guild halls to add beauty to their villages. Besides this, there are also Avalokitesvara, Lu Ban, Amitabha, etc. From the worship of a single god of hometown, to the worship of several gods based on the gods of hometown, migrants from Hunan and Hubei living in Luodai not only cover the diversified invocations of their villagers, but also show strong utility with the development of Sichuan.

It is worth mentioning that the corridor in the rear yard of Huguang Guild Hall in Luodai was once a place for migrants to deposit coffins. For those people wandering in a foreign land, occasional incidents make their fortunes unpredictable; therefore, in the past, there were places in migrant guild halls for poor migrants to deposit their coffins, and there were also free graves for those migrants who were not able to bury their relatives. Such charitable actions definitely reinforced the status of guild halls in the minds of villagers. From this, we can see the tolerance for guild hall organizations for each social class, and the self-awareness of maintaining the traditional moral principles of China. On this basis, it also participates in social management.

洛带湖广会馆建筑实测图

Building Survey Map of Huguang Guild Hall in Luodai

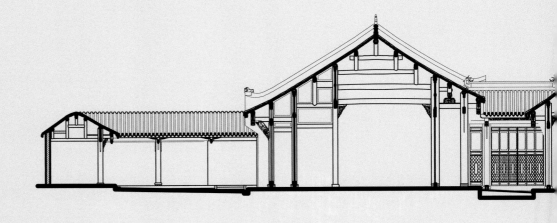

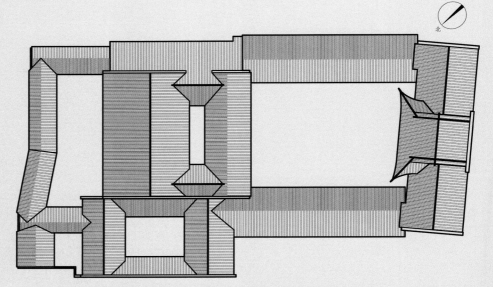

湖广会馆俯视总平面图 Overlooking General Layout of Huguang Guild Hall

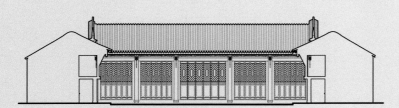

湖广会馆正殿 Ⓐ—Ⓙ 立面图
Ⓐ—Ⓙ Elevation of the Main Hall of Huguang Guild Hall

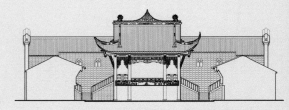

湖广会馆戏楼 ⑫—① 立面图
⑫—① Elevation of Opera Stage of Huguang Guild Hall

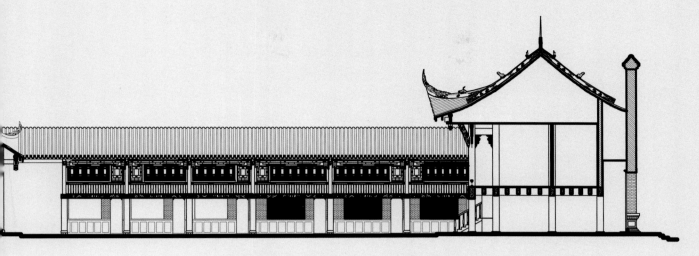

湖广会馆总纵剖面图 General Longitudinal Profile Map of Huguang Guild Hall

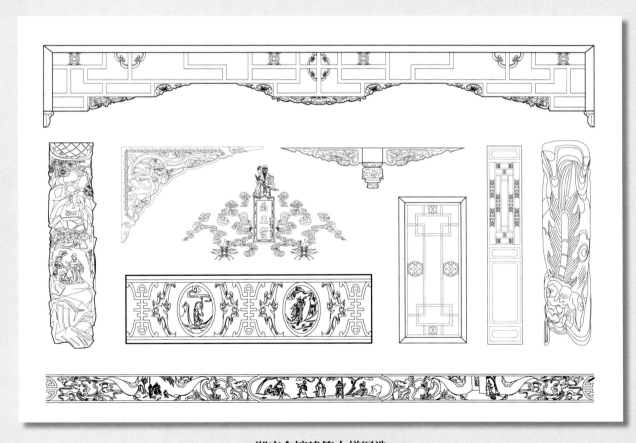

湖广会馆建筑大样图选
Building Detail Drawing of Huguang Guild Hall

　　洛带湖广会馆，是"湖广填四川"移民初期，各省移民杂居洛带、安居乐业的一个真实写照。该会馆是洛带现存始建年代最早的移民会馆，为湖广籍移民于清乾隆8年（1743年）捐资修建。

　　Huguang Guild Hall in Luodai displays how migrants from Hunan and Hubei lived in Luodai when they just settled here during the migrant wave of "Huguang Filling Sichuan". This hall, the earliest existing one, was first built in the 8th year of Emperor Qianlong's reign (1743) in Qing Dynasty by donations from migrants of Hunan and Hubei.

　　洛带湖广会馆曾供奉有湖广移民崇信的大禹宝相,所以会馆又称禹王宫。湖广移民认为,是因为大禹的治水之功,才让他们得以把足迹展向"天府之国"四川,所以湖广移民也就超越了狭隘的乡土观念,把大禹看做是他们的保护神,慕之仰之,奉之祀之。

　　The portrait of Yu the Great was enshrined in the main hall. Therefore, the hall is also known as Palace of Yu the Great. Migrants from Hunan and Hubei believe that Yu made great contributions to saving people from floods and this set the stage for their migration into Sichuan, the Land of Abundance open-minded migrants from Hunan and Hubei worship Yu and regard him as their guardian.

　　洛带湖广会馆戏楼与门楼背靠而立，戏楼底层架空，为进入会馆的门庭。整个戏楼贴金装饰，雕刻精美，富丽堂皇。戏楼隔宽阔的院坝与会馆大殿正对，其东西两侧为对称而建的一楼一底卷棚屋面厢房，厢房楼层有宽阔的虎廊，人们可边饮茶，边凭栏看戏。

　　Decorated archway and theatrical stage of Huguang Guild Hall stand back-to-back. The theatrical stage is elevated, standing at the entrance of the hall. Gorgeous theatrical stage features gold decoration and intricate carvings. A spacious courtyard stands between the theatrical stage and the main hall. In the east and west of the Guild Hall, two-storey wing-rooms with Chinese round ridge roof are built. Each wing-room is equipped with a corridor. This is a place for people to appreciate operas over a cup of tea.

三英战吕布

洛带湖广会馆戏楼的檐枋、额枋和栏板，及东西两侧厢房的护栏，均木雕有"三英战吕布"、"忠义堂"等诸如《三国演义》、《水浒传》这些脍炙人口的明清小说场景、或历史典故和民间传说。作为一种公共建筑，会馆建筑上的雕刻，除了"雕梁画栋"的建筑美学意义外，更旨在对寓外同乡起到一个劝喻、认知、教化的作用，尤其《三国演义》、《水浒传》等戏文故事的雕刻，从中反映了会馆组织对传统文化"忠、孝、礼、义"的自我规范和宣扬。

Popular fictions of Ming and Qing Dynasty, historical allusions and folklores—"Three Heroes against Lv Bu" in "The Romance of the Three Kingdoms", "Zhongyi Hall" in "Water Margin" etc.- can be found as wood carvings on eaves, architraves, and breast boards of theatrical stage and guardrails around east and west wing-rooms in Huguang Guild Hall in Luodai. As a kind of public agriculture, besides agricultural aesthetic value of "carved beams and painted rafters", carvings on guild hall architectures aim more at playing an advising, perceiving, and educational role in the life of those migrants. Especially, the carvings of stories in "The Romance of the Three Kingdoms", "Water Margin", etc. show guild hall organizations' self-discipline and high esteem for "loyalty, filial piety, manners, and righteousness".

　　洛带湖广会馆大殿后檐撑弓"狮子戏绣球",雕工精湛,形象生动。《本草纲目》记载"狮子出西域诸国",但在中国民俗文化中,狮子却因为佛教的广泛传播,而在国人心目中得到神性的升华。除了有传统意义上保平安、纳富贵的吉祥寓意外,在湖广移民的心目中,狮子更是富贵生财、子孙繁衍的象征。

　　大殿门窗花格,拙朴规矩,庄重厚实,与整个会馆的建筑风貌十分一致,从中反映出湖广移民既事雕琢,更趋于落落大方的审美特质。

　　The vivid carving "Lions Playing with Silk Strip Ball" that embodies carver's high skill can be found on the upholder of back eaves of the audience hall of Huguang Guild Hall in Luodai. According to "Compendium of Materia Medica", "Lion Comes from the Western Regions", but in Chinese folk culture, the wide spread of Buddhism gives lions a holy status in Chinese people's hearts. In addition to traditional luck implying meanings of safety, riches and honor, lions also symbolize fortune and large family in the eyes of migrants from Hubei and Hunan.

　　The doors and windows of the audience hall seem simple, unadorned, solemn and substantial, agreeing with the style and features of the guild hall, which reflects that migrants from Hubei and Hunan had a preference for natural and graceful style as well as carving.

洛带湖广会馆过厅为卷棚硬山式瓦筒屋面，大殿为单檐硬山式瓦筒屋面，二者皆为抬梁式梁架，建筑空间疏朗开阔。在大殿梁架上，有"中华民国二年癸丑阴历三月谷旦"字铭，原因是会馆曾在中华民国元年（1912年）遭遇火灾，之后于中华民国二年（1913年）重建而成。

The entryway has round ridge roof tile and the hall has a single-eave gabled roof covered with tiles. Both are designed by lift beam frames, which ensure enough space. The beam frame reads, completed in the third lunar month of the Republic of China. This hall once was wrecked by fire in the first year of Republic of China (1912). Later, in the second year of Republic of China (1913), it was rebuilt.

　　洛带湖广会馆主体建筑皆依中轴线对称布局，不过会馆左侧独辟的回廊庭院，清雅僻静，回廊戏台简朴而不失雅致，从中昭示了湖广移民对乡情和游艺的某种追求。除此之外，还需提及的是整个会馆并无下水通道，但无论下多大的雨，会馆都不会淌水漫延，如此奇迹，传为大禹保佑之故。

　　Huguang Guild Hall is arranged symmetrically along a central axis. To the left of the hall, some corridors wind through several quiet and secluded courtyards. Meanwhile, though theatrical stages are simply designed, they are still graceful in every sense. This illustrates that the attachment to hometowns cherished by migrants from Hunan and Hubei and their pursuit of recreational activity. Besides, more impressively, there is no drainage way in the hall but it is never be flooded. Truly a miracle, people believe that it is blessed by Yu.

镶嵌在洛带湖广会馆左侧庭院回廊墙上的众多碑刻，记录有当初培修会馆众多捐资者的姓名。岁月沧桑，碑刻蚀蚀，那些名字已为往事注脚的一个个模糊符号，不过至今屹立的会馆建筑，却是我们今天得以溯源追根、触之有温的史书篇章。

Many inscriptions can be found on the walls of courtyard to the left of Huguang Guild Hall in Luodai. They show the names of donators who funded the construction of the hall. As time passes by, some inscriptions are not clear. However, the hall, in itself, serves as a good textbook for us to have access to the history.

第三章
江西会馆

天地有正气,杂然赋流形……于人曰浩然,沛乎塞苍冥 ——(南宋)江西客家英杰文天祥《正气歌》

赣南客家移民与江西会馆

第三章

在明清时代，中国历史上曾有两次大的移民浪潮。一是"洪武开坎"。元末，两湖地区是元军、红巾军以及朱元璋军队厮杀的主战场，战乱导致这一地区人口锐减。朱元璋建立明朝后，曾在洪武年间下令人多地少的江西人前往两湖，这一移民浪潮一直持续到明后期，史学界称"江西填湖广"；另一次是清初，因战乱、瘟疫、天灾，曾经的天府之国一时凋敝荒芜，在清政府宽松移民垦荒政策下，以湖广籍移民为主流的各省移民，纷纷前往四川拓荒生息，史学界称"湖广填四川"。

较之于湖广移民，江西移民在"湖广填四川"之前，就行走在他乡的路上了。而大规模移民浪潮不可避免的是移民迁徙的多向性与延续性，所以明清时代的江西会馆几乎遍布全国，仅四川就有320所，其数量甚至超过湖广会馆。

洛带江西会馆地处洛带镇中街，系赣南客家人于清乾隆十八年(1753年)捐资兴建，为清代填川江西人款叙乡谊的场馆。会馆坐北朝南，主体建筑由万年台、民居府、牌坊、前中后三殿及一个小戏台构成，复四合院式，总占地2200平方米。

客家是汉民族在迁徙动荡的社会环境中，逐步形成的一支至今还保留着中原古音与中原礼仪习俗的独特民系。赣南是我国三大客家人聚居地之一。当这些赣南客家人千里迢迢初到四川时，因势单力薄，加之移民社会往往都是以同乡同籍的区分，来界定各自的利益与集团归属，移民的同籍认同也就自然大大超过了部族民系的观念，所以赣南客家人所建的洛带江西会馆，与其他地方的江西会馆相比较，其建造风格，乡神祭祀，均大同小异。

就洛带江西会馆而言，其建造手段采用了江西传统的砖木结构建造方式，先搭屋架，后建屋墙，墙倒屋不倒。其木架结构有抬梁式、井干式、穿斗式，协调搭配，灵活变化。在建筑风格上，江西会馆注重细节，如琉璃筒瓦上的兽面勾头花纹和屋脊之上的"春牛图"、"蓬山图"等图案设计精妙，不厌繁复。除此之外，洛带江西会馆在中后殿之间的天井里，还伸出一个小戏台，构思独特，环境空间布局十分完美，为四川会馆中所未曾见，从中展示了江西人独特的审美气质和赣南客家人对乡梓之情的眷念。

CHAPTER 3
Hakka migrants from South Jiangxi and Jiangxi Guild Hall

In the Ming and Qing Dynasties, there were two giant migrant waves in the Chinese history. One is called "Hongwu Kaikan". At the end of Yuan Dynasty, Hunan and Hubei were the main battlefield of Yuan Army, Hongjing Army and troops of Zhu Yuanzhang. The chaos caused by war led to sharp decrease of population in this area. After Zhu Yuanzhang established the Ming Dynasty, he had issued orders to move people from Jiangxi, where there were many people with only little land, to Hunan and Hubei. This migrant wave lasted until the later period of Ming Dynasty, and is called "Jiangxi filled Hunan and Hubei" in history circle. The other was at the beginning of Qing Dynasty. Due to wars, pestilence and natural disasters, the former Land of Abundance became destitute and deserted. In the Qing Government's policy which benefited migrants in land reclamation, migrants from each province, mainly from Hunan and Hubei, came to Sichuan to reclaim wasteland. It is called "Huguang filling Sichuan" in history circle.

When compared with migrants from Hunan and Hubei, those from Jiangxi were already on their way to foreign lands before "Huguang filling Sichuan". In the scaled migrant wave, the multi-direction and continuity of migration were unavoidable. Therefore, Jiangxi Guild Hall almost covered the whole country in Ming and Qing Dynasties; there were as many as 320 in Sichuan, the number of which even exceeded that of Huguang Guild Hall.

The Jiangxi Guild Hall in Luodai is located in Zhengzhong Street of Luodai. It was constructed with the donations from Hakka people from south Jiangxi in the 18th year of Qianlong era (1753). It was a hall for Jiangxi people filling Sichuan in Qing Dynasty to cherish their affection for hometown. The guild hall was seated in the north and facing the south. The main construction consists of Wannian Stage, Minju Residence, memorial archway, three halls of front, middle and rear, and a small drama stage. It is in the structure of a quadrangular courtyard and covers an area of 2,200 m^2.

In the chaotic social environment of Han Nationality, the Hakka people is a unique subgroup of people retaining the ancient accent and etiquette and customs of the Central Plain. South Jiangxi is one of the three biggest clusters of Hakka people in China. When these Hakka people from south Jiangxi arrived in Sichuan after a long journey, due to their weak strength and that migrant societies were always divided by their native places to identify their own interests and group affiliation, the identification between migrants from the same place naturally exceeded the concept of tribes; therefore, when compared with other Jiangxi Guild Hall, those build up by Hakka people from south Jiangxi in Luodai shared almost the same architectural styles and gods.

Speaking of Jiangxi Guild Hall in Luodai, it applied the traditional Jiangxi brick-and-wood as construction method. The frames were established at first, and then the walls were built up; the walls broke down while the houses still stand. Its wooden structure has post-and-lintel construction, log cabin construction and column and tie construction. It collocated harmoniously with strong flexibility. In architectural style, Jiangxi Guild Hall paid special attention to details: For example, the beast-face patterns on the glazed tiles, and the "Chunniu Picture", "Pengshan Picture", etc., on the ridges are all very exquisite and complicated. In addition, in Jiangxi Guild Hall in Luodai, there is a small opera stage in the courtyard between middle and rear halls. Its conception is very unique, and the environmental space layout is very perfect, which is rare among guild halls in Sichuan. It expresses the unique appreciation quality of people from Jiangxi, and the affection for hometowns of the Hakka people from southern Jiangxi.

洛带江西会馆建筑实测图
Building Survey Map of Jiangxi Guild Hall in Luodai

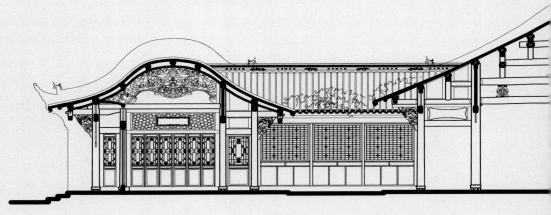

江西会馆3—3剖面图 3—3 Profile Map of Jiangxi Guild Hall

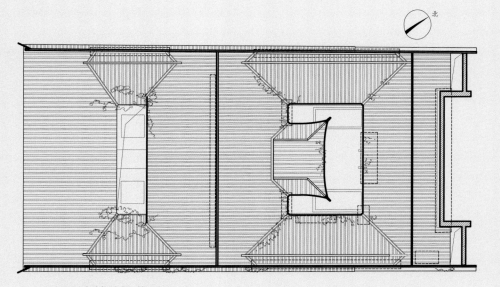

江西会馆俯视总平面图 Overlooking General Layout of Jiangxi Guild Hall

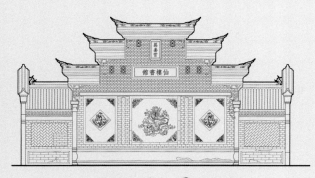

江西会馆 Ⓐ—Ⓕ 立面图
Ⓐ—Ⓕ Elevation of Jiangxi Guild Hall

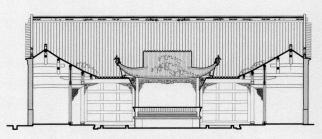

江西会馆2—2剖面图
2—2 Profile Map of Jiangxi Guild Hall

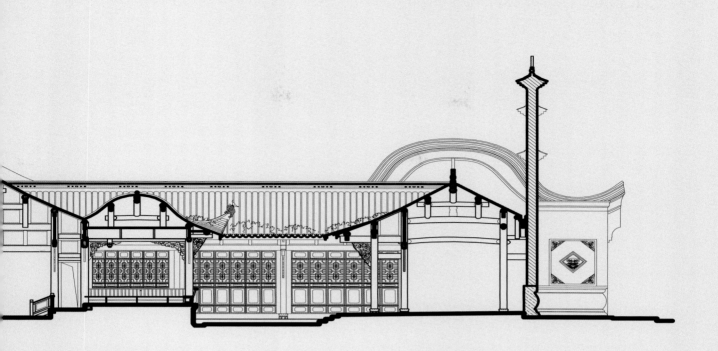

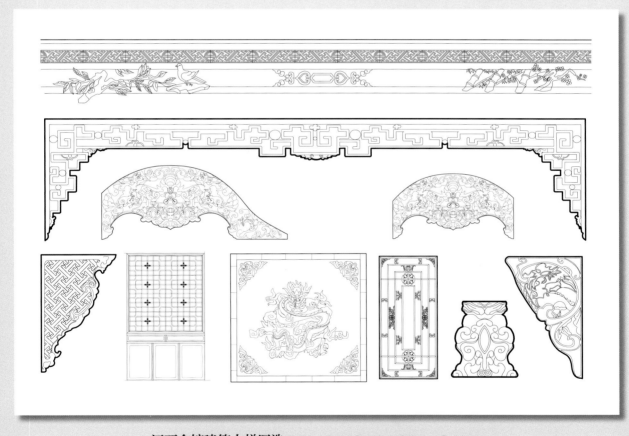

江西会馆建筑大样图选 Building Detail Drawing of Jiangxi Guild Hall

洛带江西会馆又称万寿宫，由赣南客家移民于清乾隆十八年（1753年）捐资兴建，后毁，又重建于清同治十年（1871年）。会馆坐北朝南，背街而立。在正街上，可见其独具特色的高大五花山墙。

Jiangxi Guild Hall is also called as Wanshou Palace, which was built in the 18th year of Emperor Qianglong's reign (1753) in Qing Dynasty, with donations from Hakka people from south of Jiangxi Province. It was damaged later but rebuilt in the 10th year of Emperor Tongzhi in Qing Dynasty(1871). The hall is positioned to the north and facing the south. It is at the back of a street. Unique stepped gable walls are built on the street.

洛带会馆 客地原乡 | 062
Luodai Guild Halls, Hometowns in A Foreign Land

　　从外表上看，洛带江西会馆就仿佛大户人家的宅第，及至跨进大门，朗阔的殿堂之后，是别有洞天的院落，整个建筑设计巧妙，建造精细，紧凑而华丽，既不失会馆的庄严，又充满了居家的温馨。

　　Looking from outside of the hall, Jiangxi Hall in Luodai is somewhat like the house of a well-to-do family. A courtyard will come into sight after crossing the gate and capacious hall. The magnificent architecture boasts ingenious design and compact structure. It is imposing as a guild hall and it is an inviting place as a home.

洛带江西会馆沿中轴线布局的前中后三殿及左右厢房之间，是溜口花木葱茏的天井，由此而构成这座复四合院式的庞大建筑。作为过去聚会、祭祀的重要场所，中殿是洛带江西会馆中最为轩昂的建筑单位，其正前屋面由四柱高大的木柱架梁支撑，这使大殿空间十分高耸，又兼具肃穆庄重的仪式感。

Jiangxi Guild Hall in Luodai is a large quadrangular dwelling combination, with – front, middle, and back audience halls and left and right wing-rooms constructed respectively on and along axle wire, where there are two courtyards with verdant flowers and trees surrounded by them. As an important venue for social gatherings and worship in the past, the middle audience hall is the most imposing architecture of the guild hall and its front room is supported by 4 tall wood girder erections, which make the hall being grand and solemn.

　　洛带江西会馆脊饰及小戏台飞檐套兽，都为螭的形象。传说中，龙头鱼身的螭为龙之第九子，因口润嗓粗而好吞，所以其形象多被塑为中国古建筑殿脊两端的吞脊兽，称螭吻，取其灭火消灾之意。一般说来，螭吻由龙口、鱼身和背上所插宝剑几部分组成，相传，这把宝剑为江西移民崇敬先贤许逊曾用，之所以要插上宝剑，目的是防止螭逃跑，永镇火患。螭吻是国人龙文化崇拜的延展。在洛带江西会馆前、中殿两坡垂脊上，其垂脊兽则直接采用了龙的形象，龙头高昂，大有呼风唤雨之势，其雕塑安置的目的，除了驱邪避灾的寓意外，对封护两坡瓦垄和加固垂脊也有着积极作用。

The breast carved on the cornices of the ridge and small stage of Jiangxi Guild Hall in Luodai is a legendary dragon. According to legend, it is the 9th son of a dragon, with a dragon head and a fish body. It has a smooth mouth and a gruff voice, and likes to swallow and so it is usually shaped as the swallowing ridge breast on the both end of ridges of audience halls of Chinese ancient architectures, and is called Chiwen in Chinese, meaning fire extinguishing and removing ill fortune. Generally speaking, it consists of a dragon mouth, a fish body, and a sword on its back. It is said that the sword is used by migrants from Jiangxi to worship the sage Xu Xun. The reason why a sword is inserted is to prevent the dragon from escaping and to ensure that fire can be kept away forever. Chiwen is an extension of Chinese people's worship of dragon. The vertical ridges of the two slopes of front and middle audience halls of Jiangxi Guild Hall adopted the dragon as the vertical ridge breast, with a dragon head held high and it appears that the dragon can control the force of nature. The purpose of carving such a breast is to drive out evil spirits and avoid disasters, protect tile roofs, and strengthen vertical ridges.

CHAPTER 3 赣南客家移民与江西会馆

　　曾几何时，洛带江西会馆作为学堂的厢房，书声琅琅；而会馆中、后殿间的小戏台，则不时都有江西客家人喜爱的"戈阳腔"高亢地唱出。在客家语的读书声与家乡戏曲的唱腔中，江西客家人在远离故土数千里的洛带，为自己营造出一个温暖的故乡。

　　Once upon a time, Luodai Jiangxi Guild Hall was a school wing room full of the sound of reading; while on a small theatrical stage in the rear palace of the guild hall, "Yiyang tune", a singing favored by Hakka people from Jiangxi, was performed from time to time. In the sound of Hakka reading and hometown opera singing, Jiangxi Hakka built themselves a warm place in Luodai thousands of miles away from home.

CHAPTER 3 赣南客家移民与江西会馆

　　布局精巧的洛带江西会馆，会馆与民居的建筑精华完美相融，身处其境，温馨之感油然而生，这无不得益于其建筑细节的精工细刻。仅从会馆木雕便可见一斑——前殿轩梁木雕，纹饰层层密密，扑扑飞翔的蝙蝠寓意"天降鸿福"；而后院厢房撑弓，卷草纹与锦缎纹互为穿插，直（线）曲（线）连理，动静相映，刚柔相济，庄重中又见几分妩媚，祥瑞纷呈，刻工精到，传统方胜。

　　Jiangxi Guild Hall in Luodai has exquisite layout and the agricultural essence of dwellings are integrated with it perfectly. When standing there, you can feel a sense of warmth and sweetness, which is the result of fine workmanship for the structure. The wood carvings of the guild hall give some clues. Tiers of emblazonry can be found on beams of front audience hall, flying bats on which deliver an implied meaning of "vast happiness from the heaven"; on the upholder of the back audience hall, there are carvings of floral scrolls and brocades, with straight lines and curves, dynamism and inertia, kindness and severity, charm and decency, auspicious signs, skilled carving, and Chinese traditional auspicious patterns.

洛带江西会馆主体建筑，由前中后三殿和一个小戏台以中轴线对称布局构成。其中典雅别致的小戏台，也可看着是中殿向天井延伸的亭。该亭四角立柱，藻井梁架，亭之台基兼顾为须弥式戏台，既可于台上品茗闲坐，也可作戏曲演出之地，构思精巧，为四川移民会馆所仅有。

　　The main building of Luodai Jiangxi Guild Hall consists of front, middle, rear halls, and one small theatrical stage that are symmetrically distributed along the central axis. The unique small stage can also be regarded as a pavilion extending from the middle hall towards the patio. The pavilion is supported by pillars at four corners and has a caisson ceiling; smart design makes foundation of the pavilion also a meru-style stage that can be used both as a relaxation space for tea and for opera performances, this is quite unique among migrants' guild halls in Sichuan.

洛带江西会馆后殿山墙影壁,砖砌整洁有律,灰塑巧匠精工,翼角飞翅,造型精美,不仅为古镇正体添色,也反映出江西客家人对风水的重视。

The gable and screen walls are made with bricks. Exquisite gray model art stands out. On the walls, upturned roof-ridges pointing upward are beautifully designed. All of these make the ancient town more charming. In addition, these designs also show that Hakka people from Jiangxi pay particular attention to Feng Shui.

第四章 广东会馆

月光光，秀才郎，骑白马，过莲塘
——洛带地区客家童谣

粤籍客家移民与广东会馆

第四章

"察庙之大小，即知人民之盛衰"，洛带广东会馆系全国保存最完好、规模最宏大的会馆之一，由此而见粤籍客家人在洛带一地的人口众多和影响广泛。

洛带广东会馆始建于清乾隆十一年（1746年）。会馆坐北向南，主体建筑由戏台、左右厢房、前中后殿及两个天井组成，总建筑面积达3250平方米。整体建筑体现出岭南客家楼式建筑特色——三重大殿依次排列，殿内设浑圆巨型木柱，十分庄重；沿正殿两侧梯道，可至16米高的顶楼，向北可俯瞰古镇全景，向南眺望群山，一览无遗。三殿侧外，是砖砌的封火高墙，曲线优美，起伏有致，为全川会馆之独有。

洛带广东会馆正殿称"粤王楼"，其正堂原来还供奉有头戴冕旒、身着龙袍的粤王宝相，不过后人误之为玉皇大帝，所以粤王楼又一度被误称为玉皇楼。其实，正殿底楼明间檐柱上的一则楹联就写得十分清楚——"云水苍茫，异地久栖巴子国；乡关迢递，归舟欲上粤王台"。

粤王台在广州市北越秀山上，相传为西汉时南越国缔造者赵佗所筑。在粤籍客家人的心目中，南越王赵佗是秦统一中国以来，最早成功在南方少数民族地区推行民族亲和政策的杰出政治家，是最早把中原文化和先进生产力传播岭南的伟大先驱。异乡没有粤王台，于是粤籍客家人就在洛带建起了这座"粤王楼"。

洛带广东会馆中殿次间檐柱上的楹联——"衣钵绍黄梅，浓荫遐帐，蜀岭慈云连粤岭；坛经番贝叶，宗源溥导，曲江分派接沱江"，揭示了当年洛带广东会馆的缔造者，多是从广东梅县迁徙而来的；从中也表露出他们对先贤慧能的赞誉与崇敬，和绵绵无尽的思乡情。而现代客家画家邱笑秋为广东会馆创作并悬挂于中堂的一副楹联——"叭叶子烟品西蜀土味，摆客家话温中原古音"，更是以一个细节，传神地刻画出客家人扎根四川，既溶于巴蜀文化，又以"宁卖祖宗田，不丢客家言"的祖训，顽强传承自身中原文化的精神特质。

在过去，洛带广东会馆里还供奉有庄周、老君、炳灵、南极寿星、八仙、三清九皇等神偶。在每年定期举行的粤王会里，乡党齐聚，拜祭先贤和各路神仙。祭祀之后，会馆戏台上往往有本意是酬神的戏曲演出。到后来，酬神娱人两不误，凡逢年过节，甚至同籍族人哪家有了喜庆之事，大家都会在会馆里济济一堂，在"乡谈"的嘘寒问暖中，共赏戏曲之乐。

CHAPTER 4
Hakka Migrants from Guangdong
and Guangdong Guild Hall

"The size of temple reflects the prosperity and decline of people". Guangdong Guild Hall in Luodai is one of the best preserved and the largest guild hall in China. This may be assumed that the Hakka people from Guangdong had a large population and exerted an extensive influence in Luodai.

Guangdong Guild Hall in Luodai was originally built in the 11th year(1746) of Emperor Qianlong's reign of the Qing Dynasty. With a total floor area of 3,250 m^2, the Guild Hall is with a southern exposure with the main building consists of a theatrical stage, left and right wing rooms, front, middle and rear courts, and two dooryards. The main building is characterized in typical architecture features of Hakka building in south of the five ridges. The three halls stand closely to one another with large round columns, looking very solemn. Along the staircases on both sides of the main hall located a 16m-high top floor where people can get a full view of the entire ancient town in the north and the mountains in the south. Around the three halls, there are fire-proof high walls that were built by bricks. The walls are beautiful in profile and wavy in shape, a unique style from typical guild halls found in Sichuan.

The main hall of Guangdong Guild Hall in Luodai is called "Guangdong King Tower" because of the sculpture of Guangdong King who is wearing a crown and an imperial robe. However, the Guangdong King was recognized as the Jade Emperor by mistake. This led to a false nickname of the "Jade Emperor Tower". On the first floor of the main hall lie a couplet hung on the columns, indicating that "the Hakka people come from a foreign land to Sichuan and when they feel nostalgic, they are missing Guangdong King Terrace in their hometown".

Guangdong King Terrace is located in the Yuexiu Mountain of north Guangzhou. According to legend, it was built by Zhao Tuo, the founder of Nanyue Kingdom during the Western Han Dynasty. In the heart of Hakka people from Guangdong, Zhao Tuo, the king of Nanyue Kingdom, was an excellent politician with competence in publicizing national unity policy among minor ethnic groups in southern China during the reign of Emperor Qinshihuang, a great precursor who brought central China cultures and advanced technology to the south of the five ridges and united China. The Hakka people built this Gungdong King Tower in Luodai because there was no Guangdong King Terrace in a foreign land.

The middle hall of Guangdong Guild Hall in Luodai has a couplet hung on the peripheral columns, saying "the clouds above Sichuan's mountains are connected with those above Guangdong's mountains, and the branches of Qujiang River are interlinked with those of Tuojiang River". This reveals that the founders of Guangdong Guild Hall in Luodai may come from Meixian County, Guangdong Province. It shows their worship and respect to ancient sages, as well as an endless nostalgia. In addition, modern Hakka painter, Qiu Xiaoqiu, created a couplet hung on the central hall, saying "smoke flue-cured tobacco leaves to know the folks in western Sichuan, talk with Hakka dialects to recall the histories in the Central Plain". This describes the Hakka people's cautious integration into the Bashu Culture, that they were willing to "rather sell ancestral lands than forget Hakka dialects". This reflects the spiritual characteristics of the stubborn inheritance of Central Plain culture.

In the past, Guangdong Guild Hall in Luodai enshrined sculptures such as Zhuangzhou, Laojun, Bingling, Nanji Shouxing, Baxian, and Sanqing Jiuhuang. At the annual Guangdong King Fair, villagers gathered together to worship ancient sages and gods. Drama performances followed the worship ceremony to show appreciation to their gods along with entertaining the crowds. On New Year's Day and other festivals, sometimes, even at casual parties, villagers would fill the guild hall to enjoy drama shows together, accompanying with warm greetings among each other.

洛带广东会馆建筑实测图

Building Survey Map of Guangdong Guild Hall in Luodai

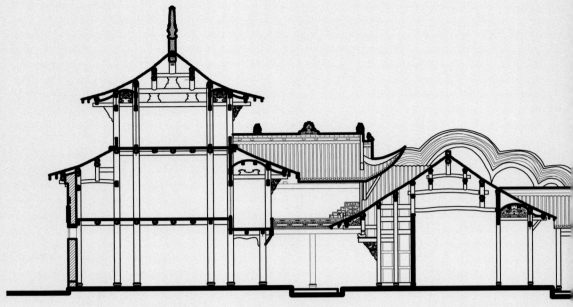

广东会馆1—1总纵剖面图 1-1 General Longitudinal Profile Map of Guangdong Guild Hall

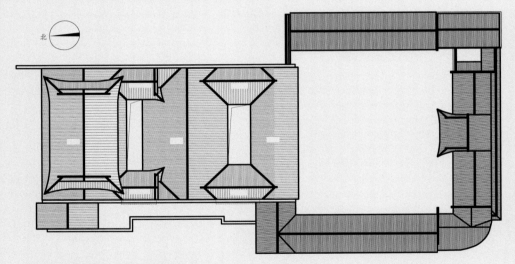

广东会馆俯视总平面图 Overlooking General Layout of Guangdong Guild Hall

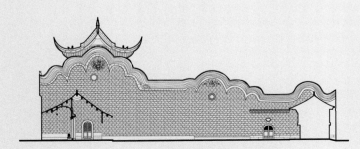

广东会馆三大殿西立面图
West Elevation of Three Palaces in Guangdong Guild Hall

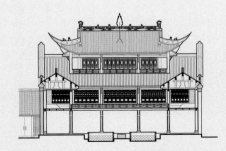

广东会馆后殿正立面、厢房4—4剖面图
4—4 Front Profile Map of Rear Court Room and Profile Map of Wing Room of Guangdong Guild Hall

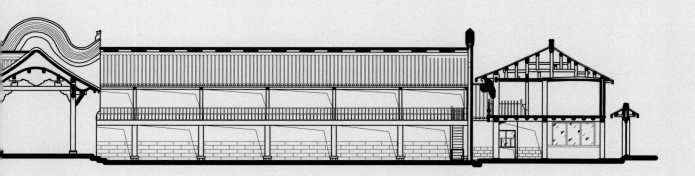

广东会馆建筑大样图选
Building Detail Drawing of Guangdong Guild Hall

洛带广东会馆又称南华宫。系广东籍客家人于清乾隆十一年（1746年）兴建。会馆占地3310平方米，坐北朝南，主体建筑由戏楼、乐楼、耳楼及前中后三殿组成，呈中轴线对称布局，复四合院结构，建造精良，造型雄伟，是目前全国保存最好、规模最宏大的会馆之一。

Luodai Guangdong Guild Hall, also known as Nanhua Hall, was built by Hakka people from Guangdong in the 11th year (1746) of Emperor Qianlong of Qing Dynasty. Facing south, the guild hall covers an area of 3,310m²; its main building consists of an theatrical stage, music building, ear building, as well as front, middle and rear halls, all evenly laid out along central axis. It is a well constructed, magnificent quadrangle structure and one of best preserved and largest guild halls in China.

洛带广东会馆后殿,又称禹王楼,是会馆中最高的建筑单元,站在其顶楼挑廊,古镇院院相接的青瓦民居,尽收眼底。

Also known as Yuewang Hall, the rear hall of Luodai Guangdong Guild Hall is the highest architectural unit of the guild hall. Standing in the corridor on top of the hall, one will be able to see every grey-tiled folk house in the ancient with their yards connected with each other.

　　在洛带四大会馆中,唯广东会馆前殿为琉璃绿瓦盖顶,后殿为黄色琉璃瓦盖顶。明清两代朝廷曾明文规定,只有皇宫、帝陵及奉旨兴建的寺庙才准许使用黄色琉璃瓦,亲王、郡王等贵族的住宅可用琉璃绿瓦外,其他建筑一律不得擅用,由此而见该会馆建筑规格之高。

　　Among the four major guild halls of Luodai, only Guangdong Guild Hall has green glazed tile roof in its front hall and yellow glazed tile roof in the rear. As expressly stipulated by the government in Ming and Qing Dynasties, only imperial palaces, mausoleums of emperors and temples built on orders of the emperor were allowed to use yellow glazed tiles, and houses of princes and county lords were allowed to use green glazed tiles; other than that, those tiles could not be randomly used, reflecting the prestigious building regulations of the guild hall.

移民会馆的基本功能除了祭祀乡神先贤外，也是会聚同乡、荟萃商贾的重要场所。以洛带广东会馆为例，其座椅排列有序的中殿正堂，就是过去会馆迎来送往、公议宴叙的礼仪性空间。

Besides being used for worshiping local gods and ancestors, migrant guild halls are also important places where home-town fellows and merchants gathered. Taking Luodai Guangdong Guild Hall for example, its middle hall where seats are orderly arranged was a place for receiving and sending off guests, having public discussions and banquets.

　　会馆虽然为民间建筑的一种，但其建筑工艺却代表了民间技艺和集体审美的最高水平。以洛带广东会馆轩廊为例，其内部吊顶曲线优美，轩梁上的花卉木雕，蝙蝠飞翔，麒麟翘首，蝴蝶蹁跹，花团锦簇，祥瑞喜庆的气氛不容言表。值得一提的是，这些精美木雕，款款相似，又款款不同，巧匠精工，让人叹为观止。

　　除此之外，洛带广东会馆的花窗也极有特色。宽大的花窗，其花格花式虽繁复而雅致，大面积的镂空设计，给予建筑内部以良好的通风和采光，较之客家民居建筑，洛带广东会馆敞亮而庄重。

　　Although guild halls are a kind of private buildings, the building craft represents the highest level of folk craftsmanship and collective aesthetics. Let's take the corridor of Guangdong Guild Hall in Luodai for example. Its suspended internal ceiling has beautiful curves; its beams are carved with flower clusters, flying bats, head-raised unicorns and dancing butterflies. It's very hard to put into words the atmosphere of auspice and joy. It is notable that these exquisite wood carvings are similar yet different. The fine workmanship is just amazing.

　　In addition, the latticed windows of Guangdong Guild Hall in Luodai are also very unique. They are broad with intricate and elegant lattice. The large hollow parts enable good ventilation and lighting inside the building. Compared with Hakka residential architecture, Guangdong Guild Hall in Luodai is spacious and solemn.

天井相隔的洛带广东会馆三殿建筑，充分体现了客家楼式建筑特色。其中后殿虎廊十分疏阔，建筑构件木雕精美，是过去粤籍客家人在祭祀、议事之余乡谈闲聚之地。

The three halls of Luodai Guangdong Guild Hall separated by patios completely represent architectural features of Hakka buildings. In particular, the corridor of rear hall is quite spacious; it has delicate wooden engravings and was used as a leisure center of Hakka people from Guangdong when they were not conducting sacrifices or having discussions.

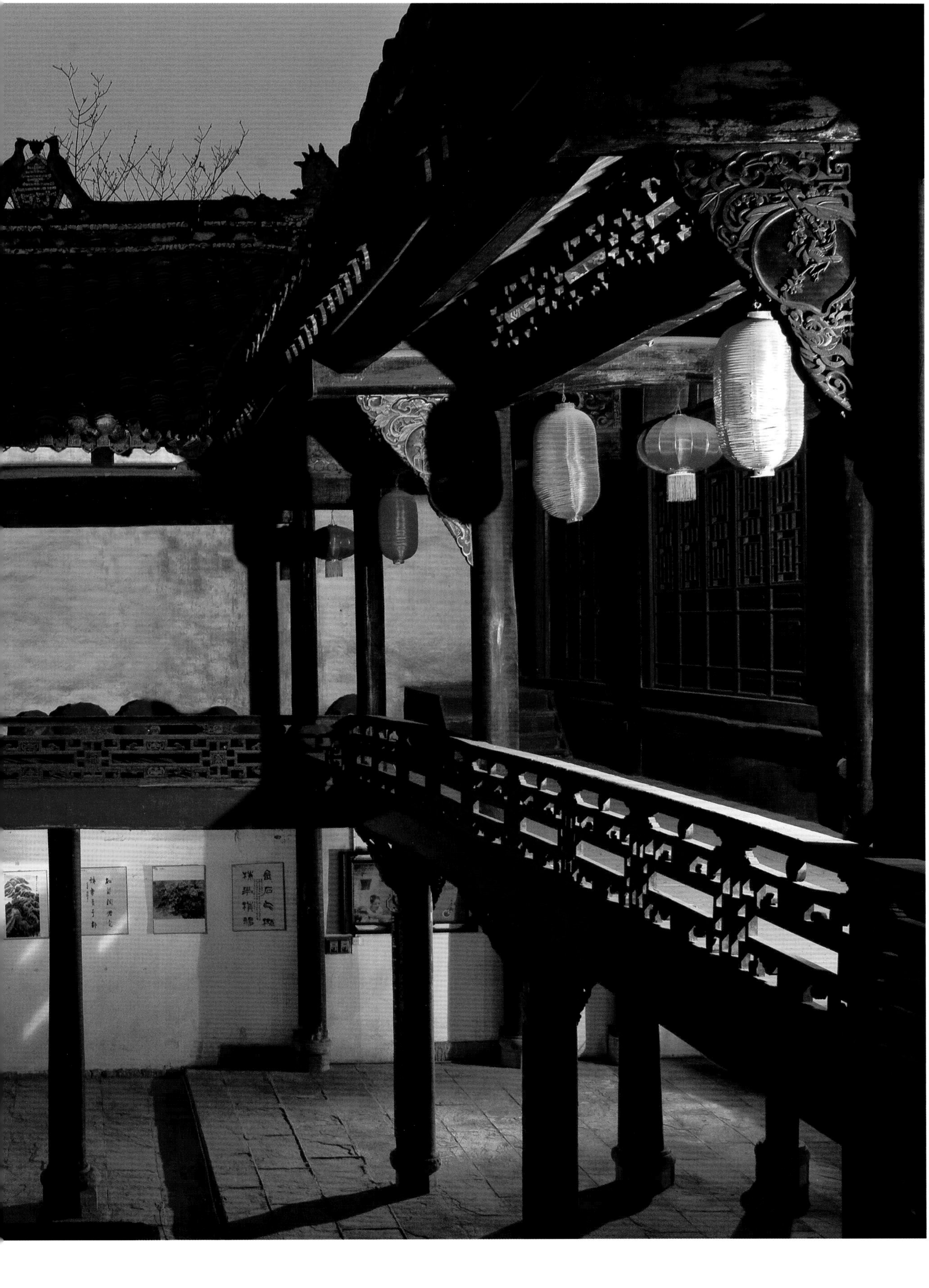

　　中国古建雕刻，随明清时期会馆等大型建筑群体的涌现，而日趋盛行。无论木雕还是砖雕、石雕，题材十分广泛，或珍禽瑞兽、山水花鸟，或历史典故、民间传说，或戏文人物、吉祥纹饰等等。就木雕而言，在洛带四大会馆中，以广东会馆撑弓上的戏文人物最具特色，《杨家将》、《三国演义》等戏文故事，构思巧妙，雕工细腻，人物形象形神兼备，栩栩如生，真可谓撑弓镂戏文，檐下多豪杰。除此之外，会馆中殿的一对双狮撑弓，线条流畅，造型生动。"狮"谐音"事"，双狮寓意"事事如意"。从中而见会馆雕刻所释放的丰富民俗文化信息。

　　Architectural carvings were becoming more and more popular in ancient China, as the springing up of large-scale architectural complex during Ming and Qing Dynasties. No matter wooden carving, brick carving or stone carving, the themes thereof are selected from a broad range of rare fowls and auspicious beasts, landscapes and flowers & birds; historical allusions, folk legends; drama characters, auspicious patterns; etc. As for the wooden carvings, among the four guild halls in Luodai, the drama characters carved on the upholders of Guangdong Guild Hall are the most distinct carvings. Based on dramatic stories such as Yang Warriors and Romance of the Three Kingdoms, the carvings are designed ingeniously and carved skillfully. The drama characters are in pursuit of verisimilitude in both form and spirit, looking like real people, which are described as "drama written on upholders and heroes living under eaves". Furthermore, in the middle hall of Guangdong Guild Hall there is a pair of upholders carved with lions on each, which are smooth in outline and vivid in appearance. In Chinese, two lions symbolize a best wish of "Everything goes well" due to the homophone. Thus it can be seen that the carvings in guild halls implies rich connotations of folk culture.

重檐歇山式黄色琉璃瓦屋面的洛带广东会馆后殿，其正脊"中堆"系瓷片和琉璃陶雕构建的一座镂空宝塔，瓷片为"岁岁（碎碎）平安"之意，宝塔则为镇邪。正脊琉璃浮雕四爪龙纹。正脊宝塔两侧，各列琉璃陶雕嘲风一尊，传其为龙之三子，生性好险又好望，作为脊兽之一，嘲风寓意威慑妖魔、清灾除祸。正脊两端为琉璃陶雕螭吻。除此之外，大殿垂脊下端面有琉璃浮雕力士，戗脊上立有琉璃雕力士。

晨钟暮鼓，岁月悠悠，殿顶上的这些脊饰在装饰和稳固屋脊、瓦垄的同时，其寓意美好的传统建筑文化，也给今天的人们带来无尽的遐想。

The back hall of Guangdong Guild Hall has double-eave gable and hip roof and yellow glazed tiles. In the center of its main ridge is a hollow pagoda made of ceramic tiles and glazed pottery, with the former standing for "peace year after year (homophones with 'broken pieces' in Chinese)" and the latter being used to repress evil. The main ridge has glazed relievo of four-claw dragons. On each side of the pagoda, there is a glazed pottery sculpture of an animal named Chaofeng, which according to legend, is the third son of a dragon and likes adventure and keeps a watchful eye. As a kind animal set on ridges, Chaofeng is intended to deter demons and keep away disasters. At each end of the main ridge, there is a glazed pottery sculpture of an animal named Chiwen. In addition, the surface of the lower end of the vertical ridges of the hall is embossed with glazed Hercules-like figures that also stand on the diagonal ridge.

Time flies with the morning bell and evening drum. These decorations on the roof of the hall not only adorn and stabilize roof ridges and tile ridges, but also bring people nowadays endless reverie with the beautiful traditional architectural culture they convey.

洛带广东会馆三殿主体建筑两边，高大的封火山墙如弓似云，曲线优美，舒展自如，错落有致。明清以前，四川建筑无论官式还是民间，封火山墙这种建筑形式都十分少见，及至"湖广填四川"，封火墙才随移民会馆的兴建，登上四川建筑的舞台。

Standing on both sides of the main building of Luodai Guangdong Guild Hall, the well-spaced fire-sealing gables look similar to the shape of bow or clouds in elegant and flexible shapes. Before Ming and Qing Dynasties, fire-sealing gables were rarely seen in Sichuan buildings, whether official or civil; and it was not until the time of "Huguang Filling Sichuan" that fire-sealing gable made its first

第五章 川北会馆
North Sichuan Guild Hall

保宁醋保宁绸……贩到成都善价求

——（清）成都竹枝词

川北会馆 晚清四川会馆的典范

第五章

地处洛带古镇上场口的川北会馆，并非洛带本物，系2000年从成都卧龙桥街异地原貌搬迁至此的。川北会馆也是洛带四大会馆中唯一一座非坐北朝南的会馆建筑，它的朝向为坐东南面西北。

北川会馆建于清同治年间（1862年~1875年），相比与洛带的湖广会馆、广东会馆、江西会馆而言，它是一座同籍会馆，是当年川北南充、西充、盐亭三县商贾、士绅，在省会成都设立的一座供同乡聚会、祭祀、下榻的场所，所以，川北会馆又名"三邑会馆"。有名的中华书局曾在川北会馆内设置分局，后来春熙路建成后，才迁往新址。

从会馆的功能讲，川北会馆本质上还是晚清时期成都市区的一个商务平台，当时很多川北的商务活动都在川北会馆进行。

清同治时代，历经200年发展的成都，曾经的"满目秽芜"之地，已为"西部的北京"。清同治十年（1871年），德国地理学家Von Richthofen 游历四川时，曾描述他对成都的印象——"中国最大的城市之一，也是最秀丽雅致的城市之一"。据相关史料，那时成都商品市场的规模化、多样化，正吸引各地客商往来贸易或坐列贩卖。早在清乾嘉时就有歌云：

郫县高烟郫筒酒，保宁醪醋保宁绸。

西来氆氇铁皮布，贩到成都善价求。

舟车交汇、商旅云集的成都，自然商号、行帮、商帮林立。出于同业相扶的利益划分，和乡情乡谊的集团归属需要，各省客商或依托移民会馆，或组织同籍商帮另建会馆。但如川北会馆这样由同省辖县人士在省会建立的会馆，并不多，由此而见当时川北商贾、士绅的活跃与重乡谊之情的乡梓观念。

川北会馆主体建筑是由大殿和乐楼构成，总体布局呈四合院式。大殿建筑构件各类花卉图案刻工精美，房顶用筒瓦覆盖，以镂空龙凤陶砖为正脊。乐楼的底高大约为大殿的一半，两相比照，构成一种参差之美。融合了中国传统会馆建筑和川北民居特色的川北会馆，因其独特精巧的建筑风格，而被誉为晚清时期四川会馆的典范。

作为川北移民在成都的这种历史文化遗存，异地重建的川北会馆丰富了洛带的会馆文化。

CHAPTER 5
North Sichuan Guild Hall
A Classic of Sichuan Guild Halls during the Late Qing Dynasty

North Sichuan Guild Hall, located in Shangchangkou, Luodai, was not constructed here originally, but was moved here from Chengdu Wolongqiao Street in 2000. The building of North Sichuan Guild Hall is the only guild hall that is not facing south but southeast among the four major guild halls in Luodai.

North Sichuan Guild Hall was constructed during the reign of Tongzhi Emperor of the Qing Dynasty (1862–1875). Compared with Huguang Guild Hall, Guangdong Guild Hall, and Jiangxi Guild Hall, North Sichuan Guild Hall is a local guild hall set up by businessmen and gentry from three counties of north Sichuan—Nanchong, Xichong, and Yanting—as a venue of gathering, ceremony, and accommodation for travelers in Chengdu, the capital city of Sichuan. Thus North Sichuan Guild Hall won a nickname as "a guild hall for three towns". The renowned Zhonghua Book Company once set up a branch in this guild hall but moved latter after the completion of Chunxi Road.

Talking in terms of guild hall's function, North Sichuan Guild Hall was naturally a business platform for north Sichuan commercial activities in downtown Chengdu during the Qing Dynasty.

During the reign of Tongzhi Emperor of the Qing Dynasty, Chengdu with a development history of 200 years was transformed from "a dirty and disorderly land" into "Beijing in the west". In the 10th year during the reign of Tongzhi Emperor of the Qing Dynasty (1871), German geographer Von Richthofen described Chengdu as "one of the largest and most beautiful cities in China". According to historical documents, goods market in Chengdu at the time attracted nationwide businessmen for trade or retail because of its product diversity. This was mentioned as early as during the reign of Qianlong and Jiaqing Emperor of the Qing Dynasty:

Pixian County was famous for its tobacco and low wine, while Baoning was known for its vinegar and silk.

Pulu fabric from the west is as strong as iron, which can be sold to Chengdu at high prices.

As a water and land traffic hub and gathering place for businessmen, Chengdu naturally had a lot of trade firms, trade associations, and business groups. Businessmen from different provinces built new guild halls with the supports of immigrant guild hall or through organizing fellow business groups, thus to get interests of mutual supports among the industries and the sense of belonging. But guild halls constructed in the provincial capital by home fellows like the North Sichuan Guild Hall are not seen frequently, which shows that at that time, businessmen and gentry from north Sichuan were more active and appeared to show more connection to their hometown than businessmen from other parts of Sichuan.

The main structure of North Sichuan Guild Hall consists of an audience hall and a building for drama performance during worship for Zhuge Liang on Tomb-sweeping Day, generally in the form of quadrangle dwelling. In the audience hall, there are a variety of flower pictures of high carving skill, roof covered by imbrex. The main ridge is made of hollow-out dragon and phoenix earthenware bricks. The height of ancient performance stage is half of that of the hall, appearing unsymmetrical yet beautiful. North Sichuan Guild Hall, an integration of Chinese traditional guild halls and unique folk houses in north Sichuan, is honored as an example of guild halls during the late Qing Dynasty because of its unique and exquisite architectural style.

North Sichuan Guild Hall, a historical cultural heritage built by immigrants from north Sichuan in Chengdu was moved to and reconstructed in Luodai, thus enriched culture of Luodai guild halls.

洛带川北会馆建筑实测图
Building Survey Map of North Sichuan Guild Hall in Luodai

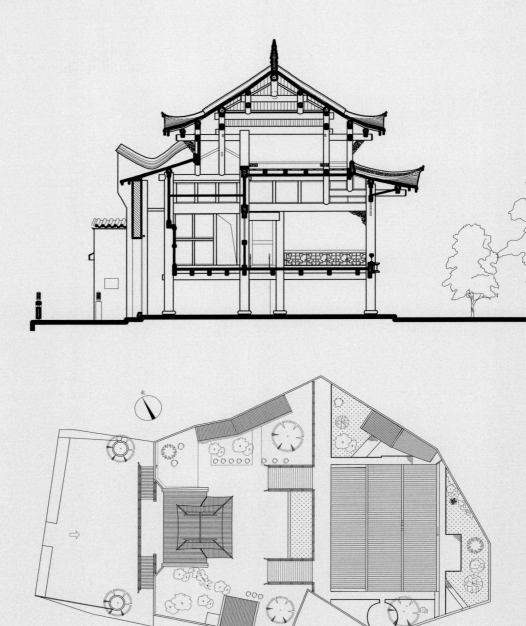

川北会馆俯视总平面图 Overlooking General Layout of North Sichuan Guild Hall

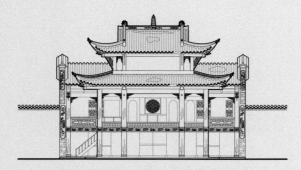

川北会馆山门、戏楼东立面图
East Elevation of the Gate and Opera Stage of North Sichuan Guild Hall

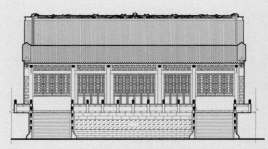

川北会馆正殿立面图
Elevation of the Main Hall of North Sichuan Guild Hall

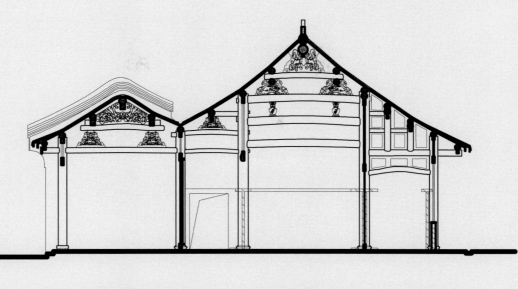

川北会馆总纵剖面图 General Longitudinal Profile Map of North Sichuan Guild Hall

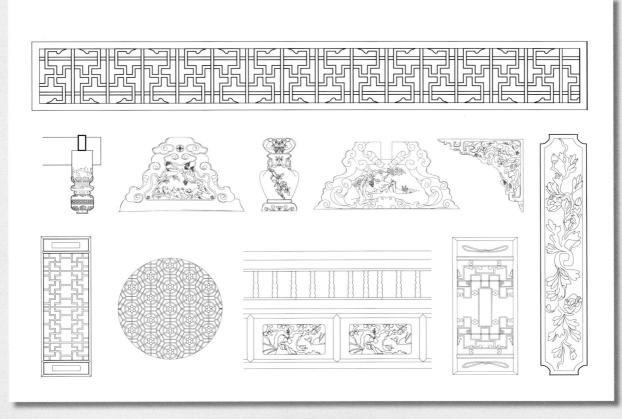

川北会馆建筑大样图选
Building Detail Drawing of North Sichuan Guild Hall

地处洛带古镇上场口的川北会馆，系2000年从成都卧龙桥街异地原貌搬迁至此的。迁入前，破败不堪，迁建后，整个会馆修葺一新，基本恢复原样。川北会馆始建于清同治年间，其主要功能是清末川北客商在成都的一个重要商务平台。川北会馆的迁入，丰富了洛带的会馆文化。

North Sichuan Guild Hall was relocated to Shangchangkou, Luodai from Chengdu Wolongqiao Street in 2000. The formerly dilapidated building of North Sichuan Guild Hall was generally restored to its original appearance after being moved. North Sichuan Guild Hall was established during the reign of Tongzhi in the Qing Dynasty. In terms of guild hall's function, North Sichuan Guild Hall naturally was mainly a business platform for north Sichuan's commercial activities in downtown Chengdu in the late Qing Dynasty. North Sichuan Guild Hall enriched the culture of guild halls in Luodai.

洛带川北会馆万年台，是成都地区会馆建筑中唯一保存较为完整的古戏楼。该戏楼楼高三层，通高13米，面阔三间，抬梁梁架，雕花门窗；一楼为会馆山门进入大殿的过道，二楼为戏台，二楼与三楼相隔，仅留一孔需借木梯方可抵达；戏楼采用重檐歇山式素瓦筒屋面，屋檐起翘，反宇向阳，减少了屋檐对光线的遮挡，加之戏台适当的高度，保证了观众不会受阳光阴影影响观戏的视野。

The Wannian Stage in North Sichuan Guild Hall in Luodai is the only completely-preserved ancient theater building in guild halls in Chengdu. The three-storey theater building is 13 meters high with post and lintel roof framing and latticed doors and windows. It has four columns in the front. The first floor is the aisle connecting gate of the guild hall to the main hall. The second floor is the stage. The second floor and the third floor are separated, leaving only a hole that one can pass through with a wooden ladder. The theater building has double-eave gable and hip roof with plain semicircle-shaped tiles. The eaves are warped towards the sky, which helps to reduce the light shielded by the eaves and, together with the proper height of the stage, ensures that the view of audience is not affected by shadows.

　　洛带川北会馆的建造，巧妙地利用了地形的自然坡度，使整个会馆形成极具特色的竖向空间。从戏楼架空的山门进入，拾级而上就是会馆大殿。这样的高差，加之大殿也分为开敞式和封闭式两大空间，这使大殿开敞的空间和殿前过道，成为主要的观戏及聚会区域。

The construction of North Sichuan Guild Hall in Luodai makes clever use of the natural slope of the terrain, so that the whole hall forms very unique vertical space. Enter from the overhead gate of the theatrical stage and go up the stairs, and you'll see the audience hall. The height difference, together with the division of the hall into open and closed space, makes the open space and the passage in front of the hall major areas for enjoying dramas and having gatherings.

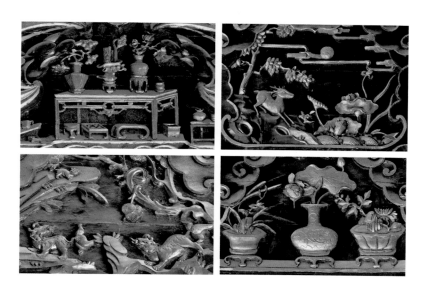

较之其他公共建筑不同的是，会馆建筑在传统祥瑞纹饰的组合运用上，更多地体现了民间大众的审美情趣。以洛带川北会馆大殿轩梁及梁架驼峰上的精美木雕为例，这些图案多以麒麟、鹿、猴、马、荷、梅以及香炉、宝瓶等为表现对象，所谓"麒麟送子"、"福禄（鹿）连（莲）年"、"平（瓶）安和（荷）美（梅）"，马（马）上封侯（猴）长命百岁"等等，将上层的祥瑞思想转变为"福寿禄喜财"世俗化的吉祥观念。

Compared with other public buildings, guild hall buildings give more expression to the aesthetic taste of common people in terms of combination and use of traditional auspicious carvings and decorations. Let's take the exquisite wood carvings on the beams and the camel-hump shaped support for the beam frame in the main hall of North Sichuan Guild Hall in Luodai for example. Mostly the patterns are unicorns, deer, monkeys, horses, lotus, wintersweet, incense burners, treasure bottles, etc., all of which are homophones for auspice-praying phrases. In this way the upper class's prayer for auspice is turned into earthly wish for blessing, longevity, emolument, happiness and wealth by common people.

洛带川北会馆的建造，充分结合了川北民居简洁明快、朴素淡雅的格调。其大殿悬山式建筑单元，采用密撩穿斗式构架，林木掩映之下，大殿木柱粉墙，青瓦黛脊，别有风味。

The simplicity and elegance of dwellings in north Sichuan are fully integrated into North Sichuan Guild Hall in Luodai. The overhanging gable roofed building units of its main hall are column and tie construction. Surrounded by trees, the main hall looks distinctively attractive with its wooden columns, whitewashed walls, grey tiles and black ridges.

CHAPTER 5
川北会馆　晚清四川会馆的典范

　　洛带川北会馆主体建筑现仅存乐楼（万年台）和大殿两部分。戏楼戏台四角立柱，三面开放，保证了戏台表演的充足光线。作为过去酬神唱戏的一种表演空间，戏楼面向大殿，并与大殿同处一条中轴线上，以此形成一种一一对应的关系。戏楼曲折多姿，典雅大方；悬山式和硬山卷棚式组合屋顶的大殿，庄重朴实，又不失精巧细腻，其独特的建筑风格，被誉为晚晴时期四川会馆建筑的典范。

　　Among the main buildings of North Sichuan Guild Hall in Luodai only the theater building (Wannian Stage) and the bobby remain. The stage has a column in each of the four corners. It is open on three sides, which ensures adequate lighting for performances on the stage. As a place for thanking gods by singing operas in the past, the theater building faces the main hall and has the same central axis as the main hall, forming a one-to-one corresponding relationship. The theater building is intricate and elegant. The combination of overhanging gable roof and round-ridge flush gable roof makes the main hall solemn and simple, yet sophisticated and delicate. With its unique architectural style, it is honored as an example of guild halls from during the late Qing Dynasty.

早在夏商周时代，古人就以乐舞来敬神祭祖，歌功颂德，而会馆作为一种祭祀的场所，酬神唱戏的戏楼由此而成为会馆建筑的重要组成部分。建造精美的洛带川北会馆戏楼，既反映了川北移民爱好戏曲的独特文化气质，也反映了其时成都戏剧文化的繁荣。

People have danced as a way to worship gods and ancestors and to sing the praises of someone from as early as Xia, Shang and Zhou Dynasties. As places for offering sacrifices, guild halls include theatrical stages, where operas were sung for gods, as their integral parts. Beautifully constructed theatrical stage in Luodai North Sichuan Guild Hall reflects the unique preference of migrants from North Sichuan for operas and the prosperity of operas in Chengdu at that time.

洛带川北会馆万年台墙体上，花鸟浮雕灰塑十分显眼，其主要内容为梅花、兰草、牡丹和仙桃。在中国传统文化中，梅花寓意美好，兰花寓意高洁，牡丹寓意富贵，仙桃寓意长寿。这些为广大民众普遍喜爱的纹饰，因塑造技法的不同，难掩会馆这种民间自发建造的公共建筑形式，因针对特定人群而设所具有的独特审美个性。而这种个性，也充分体现了会馆建筑对空间转换的需求。

On the wall of Wannian Stage in North Sichuan Guild Hall in Luodai, relievo and lime sculptures, mostly including wintersweet, orchid, peony and peach, are very attractive. In traditional Chinese culture, wintersweet means wonderfulness, orchid: loftiness, peony: richness, and peach: longevity. Although these ornamentations are favored by the general public, the different shaping techniques conceal the unique aesthetics of guild halls, a kind of building voluntarily built by civil societies for specific groups of people. The unique aesthetics give full play to the demand of guild hall buildings for spatial switching.

Worship in Guild Hall

第六章
会馆祀奉

大地回春，万象更新……虔诚祭祀，感恩追思 ——洛带会馆清明公祭祭词

会馆祀奉 乡情联络的纽带

第六章

在异乡"安土重建"的艰难岁月里，移民们本能地就会想到，可以攀附的力量，除了身边寥寥无几的亲朋好友外，更多的还是同迁一地的同籍乡人。于是，在乡土的旗帜下，他们自然而然地结成一种可叙乡愁、患难共济的松散组织，并伴随会馆的建立，一同成为移民们力图与乡梓文化保持紧密联系的一种载体。

从礼制建筑的角度讲，会馆脱胎于宗祠。对应于宗祠的血缘关系，会馆是乡情乡谊联系的纽带。可以说，一座移民会馆就是一座放大了的宗祠，是不同时空下扩大、再组织化的精神礼仪空间。宗祠的首要功能是祭祀祖先，而会馆祀奉的，则是同籍移民共同信仰的乡土神。从祖先祭祀到乡土神祀奉，这种血缘宗亲观念的扩大与演变，标志着中国平民社会的兴起与确定。

在中国传统社会里，乡土神祇既是国人地域精神、文化气质的一个象征，也是一方风土共同道德价值观的体现。对于寓外的同乡移民而言，乡土神祇不仅是他们最易认同的集体象征，更是他们维系情感的纽带。所以，乡土神灵祀奉，是移民会馆的首要功能，在特定历史环境中，它是移民社会的熔冶，也是同籍移民道德、文化倾向的标榜和规范。

移民的籍贯不同，祀奉的神祇自然也就有所不同。在洛带，江西会馆祀奉的是感天大帝许真君，惟愿其恩施长久，所以江西会馆又名万寿宫。据史料记载，许真君姓许名逊，是晋代著名的道士，汝南人（今河南家许昌），家住南昌，晋太康元年（280年）曾出任四川旌阳县令，一生为民除害，匡世济民。在晋代，神仙故事尤其泛滥，许逊这位在民众心目中有着良好印象的历史人物，自然也就在百姓口口相传的"先进事迹"中，逐渐被神化为一位得道的神仙了。

传说中，许逊曾"斩蛟除害"，为江西人避免了一场大的灾难，所以江西人都崇奉许逊，称他为"江西福主"。许逊信仰始于唐代，到宋朝被朝廷所倡导，信奉道教的宋徽宗封其为"神功妙济真君"，一时间，道观、道家诵经修道的坛靖，供奉许真君处，不计其数。当江西移民辗转千里来到他们保护神曾经为官的四川时，基于广泛的信仰基础，自然也就将其作为地方的精神代表，供奉于会馆之中了。

广东会馆祀奉的是佛教高僧六祖惠能。相传北魏年间，达摩从印度来到中国，提出中国化佛教禅宗的修行方法。达摩把他的这一禅法传给慧可，慧可又传给僧璨，然后传道信、弘忍。弘忍之后分南北二系——神秀在北方传法，建立北宗；惠能在南方传法，建立南宗。北宗神秀不久渐趋衰落，而惠能的南宗在朝廷支持下，成为中国佛教的主流，惠能也因此成为禅宗实际上的创始人。由于从达摩到惠能经过六代，故佛教中尊达摩为"初祖"，惠能为"六祖"。

史料记载，惠能（638年~713年）俗姓卢氏，祖籍河北燕山（今涿州），唐时随父流放岭南新州（今广东新兴）。在广东韶关曹溪河畔的南华寺，惠能在此传授佛法37年，法眼宗远传世界各地，所以粤籍人士皆视其为乡土神，而鉴于南华寺是六祖慧能弘法的道场，因此广东会馆又称南华宫。在今天，作为中国历史上最有影响的思想家之一，惠能与孔子、老子并列为"东方三圣人"，其思想所包含的哲理和智慧，至今也给人以有益的启迪，并越来越受到广泛的关注。

正如民间普遍供奉多神一样，广东会馆除了祀奉六祖惠能，也祀奉先贤粤王赵佗。

与江西会馆、广东会馆不同的是，湖广会馆祀奉的并非荆楚乡神，而是大禹，原因是"禹王疏九州使民得陆处"，加之过去的两湖地区，连年水患，祀奉大禹，也有借大禹之威镇水之意，所以，湖广会馆也称"禹王宫"。

禹是中国传说时代与尧、舜齐名的贤圣帝王，他以毕生精力治理洪水，又划定中国九州，后世之人为表达对禹的感激之情，尊称他为"大禹"，即"伟大的禹"。据杨雄《蜀本纪》载，"禹本汶川郡广柔县人也，生于石纽（四川汶川）"。他曾跨越蜀地，治理江河自夔门以下并疏导沔水（今汉江）进入长江，平生之业，遍及荆楚。所以自古以来，两湖百姓都敬之为神。据《湖北通志》载，其时汉阳等十九县均建有禹王庙。

在艰险的迁徙路上，当湖广移民入夔门，历丰都，面对种种险滩恶水时，他们唯有寄望故土禹王庙祀奉禹王的护佑，而当平安入蜀后，禹王也就自然被湖广移民看做是他们的保护神，并供奉于会馆高堂之上，所谓"各别其郡，私其神，以祠庙分籍贯"，成为有别于别地移民的标识。

可见，会馆祀奉的神灵，都是传统文化美德的化身，这对于一度脱离了社会管理的寓外移民而言，既发挥规范人心的作用，也使他们能够以一种符合当时社会规范的道德水准，迅速整合族群，融入当地社会。在一定程度上讲，会馆的神灵崇拜，是会馆得以实现社会整合的精神中枢。

CHAPTER 6
Worship in Guild Halls
Links Connecting Affections for Hometown

In the hard times of reconstructions in a strange place, besides rare relatives and friends, migrants would naturally regard their hometown fellows, who have moved to the same place, as a power they can attach to. Therefore, in the name of hometown, they form a loose organization naturally, in which they can express their nostalgia and help each other in tribulations. With the establishment of guild hall, it becomes a kind of carrier, through which the migrants can stay in close touch with the culture of their hometowns.

From the perspective of etiquette construction, guild hall comes from ancestral hall. Corresponding to the genetic connection of ancestral hall, guild hall is the link of provincialism. To some degree, a guild hall of migrants is an expanded ancestral hall. It is a spiritual etiquette space which is expanded and reorganized in different times and spaces. The prior function of ancestral hall is to worship ancestors, while the god of hometown, in which migrants from the same hometown believe, is consecrated in guild hall. From worship of ancestors to gods of hometowns, the expansion and evolution of this clan concept marks the rise and establishment of Chinese civilian society.

In the traditional society of China, god of hometown is not only a symbol of regional spirit and cultural temperament for Chinese people, but also a reflection of the common moral value of one place. For migrants from the same hometown in a foreign land, god of hometown is not only the symbol of the group, which they are the most likely to approve, but also the link for them to maintain their emotions. Therefore, to consecrate gods of hometown is the main function of immigrant guild hall. In specific historic environment, it is the smelting-furnace of migrant societies, and also the model and standard of some moral and cultural trends of migrants from the same place.

As migrants come from different places, they believe in different gods. In Luodai, Jiangxi Guild Hall worships Gantian Emperor Xu Zhenjun, hoping that he can bring fortune to them for a long time; therefore, Jiangxi Guild Hall is also called as Wanshou Palace. According to historical materials, the name of Xu Zhenjun is Xu Xun, and he was a Taoist priest in Jin Dynasty. He was born in Runan (currently Xuchang of Henan), but resided in Nanchang. In the first year of Taikang Emperor of Jin (208), he was appointed as the prefect of Jingyang County. He got rid of evils for the people and relieved the people in his lifetime. In Jin Dynasty, fairy tales overflowed everywhere. Xu Xun, who enjoyed high reputations in the public, became a typical character passing from mouth to mouth among the public. And he was deified as an immortal gradually.

According to legends, Xu Xun had killed a flood dragon and saved the people of Jiangxi from a disaster; therefore, the people of Jiangxi worship Xu Xun and honor him as "Mascot of Jiangxi". The belief in Xu Xun started from Tang Dynasty, and was promoted by the government in Song Dynasty. Hui Emperor of Song, who believed in Taoism, conferred the title "Man with Magic Power and Kindness" on him. In a short period of time, many Taoist temples and forums, in which people recited classics of Taoism, consecrated Xu Zhenjun. When migrants of Jiangxi moved to Sichuan, where their gods had served as an official, due to the extensive basis of belief, they naturally regarded him as the representative of local spirit, and consecrated him in guild halls.

Guild halls of Guangdong consecrate Huineng, an eminent Buddhist monk. According to legends, in the Northern Wei Dynasty, Bodhidharma came to China from India, and proposed the cultivation method of Zen Buddhism in Chinese style. Bodhidharma passed this method to Huike, and Huike passed it to Sengcan, and then to Daoxin and Hongren. Afterwards, Hongren was divided into north and south divisions: Shengxiu imparted the dharma in the north and established Beizong; Huineng imparted the dharma in the south and established Nanzong. Beizong of Shenxiu declined gradually before long, while Nanzong of Huineng turned to be the mainstream Buddhism in China with the support from the government. Actually, Huineng became the founder of Zen. There were six generations between Bodhidharma and Huineng, so in Buddhism, Bodhidharma was honored as "Initial Master", and Huineng as "the Sixth Master".

According to historical materials, the original family name of Huineng (638-713) is Lu. His ancestral home is Yanshan of Hebei (currently Zhuozhou), and in Tang Dynasty, he was exiled to Xinzhou at the south of the Five Ridges (currently Xinxing of Guangdong). Huineng had imparted the dharma in Nanhua Temple by the Caoxi River in Shaoguan of Guangdong for 37 years. His dharma has spread across the world, so Cantonese regard him as the god from their hometown. As Nanhua Temple is the forum where Huineng imparted dharma, guild hall of Guangdong is also called Nanhua Palace. At present, as one of the most influential ideologists in the history of China, Huineng is listed as the "Three Wise Men of the East", accompanied by Confucius and Lao Tse. The philosophy and wisdom contained in his thoughts still inspire people of modern time, and receives more and more attention.

As many gods are consecrated in the public, besides Huineng, guild hall of Guangdong also consecrate Zhao Tuo, King of Guangdong.

Different from Jiangxi and Guangdong guild halls, Huguang Guild Hall consecrates Yu the Great instead of gods of their hometowns. It is because that "Yu dredged the rivers and saved the people from floods", and in the past, Hunan and Hubei suffered from floods every year and tended to believe in Yu the Great. It also shares the meaning of dominating the water with Yu the Great. So, the Huguang Guild Hall of is also called "Palace of Yu the Great".

In the legendary era of China, Yu shared the reputation as judicious emperors with Yao and Shun. He spared no efforts during his life to manage floods, and then delimitate the nine districts of China. In order to express their gratitude towards Yu, the descendants address him respectfully as "Yu the Great". According to Yang Xiong's Shubenji, "Yu is a native of Guangrou County of Wenchuan Shire County, and was born in Shiniu (Wenchuan of Sichuan)". He crossed ancient Sichuan to manage the river below Kuimen and guide the Mianshui River (currently Hanjiang River) to the Yangtze River. His achievement in his lifetime covered the whole Hubei and Hunan. Therefore, people from Hubei and Hunan have honored him as a god since the ancient times. According to Hubei Tongzhi, there were Temples of Yu in 19 counties, including Hanyang.

In the perilous journey, when migrants from Huguang passed through Kuimen and Fengdu and confronted many dangerous shoals and floods, they could only entrust their hopes on Yu the Great, who was consecrated in the Yu Temples in their hometown. When they entered Sichuan safely, Yu the Great was naturally regarded as the guardian of migrants from Huguang, and was consecrated in the guild halls. This is exactly what is called "different counties have different gods; therefore, guild halls differs from native places". Guild halls became an identification of migrants, which may differ from one another.

Obviously, the gods consecrated in guild halls are reflections of traditional culture and morality. As for the migrants in other places who have been cut off from social management, it not only standardizes the will of people, but also provides convenience for migrants to integrate their groups and blend into local societies with a morality level which conforms to the current social standard. To some degree, the worship of gods in guild halls is the spiritual pivot for the guild halls to realize social integration.

在移民的心目中，会馆是"庙"，是他们祀奉原乡神祇和先贤的地方。而移民籍贯不同，其所供奉的神祇自不相同。在洛带，湖广会馆祀大禹，江西会馆祀许逊，广东会馆祀六祖惠能和粤王赵佗，他们皆为传统文化美德的化身，这对于寓外同籍乡人而言，既起到延续原乡礼俗的教诲作用，也有利于同籍组织在异乡的有效整合。图为洛带广东会馆隆重肃然的岁时祭祀。

In the eyes of the migrants, guild halls stood for "temples", where they worshiped their old gods and ancestors. And in Luodai, there were different migrants worshiping different deities. In Huguang Guild Hall, people would worship Yu the Great; people at Jiangxi Guild Hall would worship Xu Xun; Guangdong Guild Hall would worship The Sixth Patriarch Hui Neng and Zhao Tuo, King of Nanyue...all of those deities embodied the virtues in traditional cultures. For the home-town fellows, they had not only passed on tuition of traditional customs, but also benefitted the gathering of fellow people in a foreign land. The picture shows the solemn annual ritual in Luodai Guangdong Guild Hall.

　　洛带客家龙舞的龙舞者，都来自祖籍江西的一支刘姓龙舞世家。这个家族世代相传的龙舞绝技，至今仍较为完整地保留着中国古代龙舞的古朴程式。每次舞龙之前，龙队都要在洛带江西会馆举行隆重的会馆祭献，并歃血祭龙，以通神祭灵，神附龙体，其中蕴含了江西籍客家人趋吉避凶、追求幸福的美好愿望。

　　The dancers of Luodai Hakka Dragon Dance are all from a Liu family well-known for dragon dance with origins in Jiangxi province. The unique dragon-dancing skill possessed by the family still maintains a quite complete pattern of ancient Chinese dragon dance. Before every performance, the dancers would host a grand ritual at Luodai Jiangxi Guild Hall, during which they would take a blood oath to offer the dragon to god, so that the god would bestow blessings upon the dancers, and also the dragon. The ritual demonstrates the Hakka people's wish to pursue good fortune and avoid disaster, as well as for well-being.

在洛带，会馆祭祀以清明公祭最为隆重。其时，会馆幡旗飘扬，香烟缭绕，随着一阵礼炮声响，主祭官带领陪祭，走向祭台，先行"三献"（帛、爵、三牲）之礼，之后宣读祭文，行"三祭"（水祭、土祭、火祭）之礼，再敬"三娱"（舞彩龙、跳傩舞、弹古乐）……气氛庄严隆重，肃穆热烈，充分表达了客家移民对故土、先祖的崇敬感恩之情。

Among Luodai's guild hall ceremonies, the public memorial ceremony during Qingming is the most magnificent one. During that, there would be flags waving and joss sticks burning. After several gun salutes, the officiant would lead the co-officiants towards the altar to perform the "three offering rituals" (yellow silk, tripod ritual wine vessel and sacrificial pig/cow/goat); then read the oration, followed by the "three sacrificial rituals" (of water, earth and fire) and finally the "three entertaining rituals" (dragon dance, Nuo dance and ancient musical performance)…the whole ceremony is solemn, magnificent and enthusiastic, fully showing Hakka migrants' worship and respect for their homeland and ancestors.

在洛带，客家移民会馆的岁时祭献，完全依照古时皇家的蒸尝规制施行。古时，皇家有四时之祭：春祭曰礿，夏祭曰禘，秋祭曰尝，冬祭曰蒸，之后概称为"蒸尝"，其规格分"祭"、"酹"两种，按客家风俗，用猪、牛、羊"大牲"行三献礼谓之"祭"；用鸡、鱼、猪肉"小牲"，以酒祭地称为"酹"。待祭祀结束，神灵们已经享用的"大牲""小牲"自然就成了同乡聚餐的食材，所以，一顿乡情联谊的蒸尝宴席，往往成为会馆祭祀的压卷之作。

In Luodai, the annual ritual of Hakka migrant guild halls is totally in accordance with the ancient Hakka royal rules. In ancient times, the Hakka royal had four rituals for the four seasons, i.e. Si Sok in spring, Si Di in summer, Song in autumn and Ziin in winter, which were jointly called "Ziin Song". Based on the scale of the rituals, they had different names meaning "sacrifice" and "libation" in Hakka custom. The three offering ritual with "large cattle" such as pig, cow and goat was called "sacrifice"; and that with "small cattle" such as chicken, fish and pork as well as wine was called "libation". After the ritual, the cattle enjoyed by gods would become a feast for the attendees, and hence the closure of a guild hall ritual.

　　会馆神灵是移民会馆存在的前提,所以,在移民会馆的建筑设置中,庙宇与殿堂等祭祀礼仪性建筑,是一座完整移民会馆的首要构成。而从祭祀礼仪的角度讲,会馆是放大了的宗祠,只不过宗祠是同一血缘的族人祭祀祖先的场所,而会馆则是同籍乡人敬奉乡神、先贤的地方。从祠堂祭祖到会馆敬贤,表明中国古代士绅文化中的忠、孝、节、义,正悄然地融入庶民文化之中。

　　Since deities were the prerequisite for the existence of guild halls, temples, palace halls and other ceremonial buildings therefore became the primary structure of a complete migrant guild hall. In terms of rituals and ceremonies, a guild hall is actually an enlarged ancestral temple: while an ancestral temple was for a lineage to worship ancestors, a guild hall was for home-town fellows to worship gods and ancestors. Both of them represented that the virtues of loyalty, piety, chastity and righteousness in the ancient Chinese gentry's culture were blending into the fold culture quietly.

第七章
会馆楹联

Couplets Hung on the Columns of Guild Halls

华筒俱成桑梓地，乡音无改，新增天府冠裳 ——洛带广东会馆楹联语

会馆楹联 解读移民心灵密码的钥匙

第七章

洛带的几大会馆，其建筑的布局，无一不是中轴线对称布局——祭祀的大殿，议事的大厅，或精巧或宏大的戏台，在中轴线上组成两进或数进院落的建筑序列，待客的厢房等建筑则对列于中轴线两旁。

除此之外，洛带土生土长的湖广会馆、江西会馆、广东会馆，则无一例外地大致坐北朝南向，除了这些建筑都有面向移民原乡的意味外，先哲老子所言"万物负阴而抱阳，冲气以为和"，应该是建造者们最初朴实的想法。在国人传统的风水观念中，北方为水，是阴；南方为火，是阳。因此坐北朝南、负阴抱阳便成了中国大多数建筑的布局方式。

而阴阳二元观念，不仅是中国古人的基本世界观，更是中国人文文化的基础。"天地之道，有左有右，有阴有阳"，如此深入国人骨髓的阴阳观念，表现在民族心理上，重要的特征之一，就是对"对称之美"的执着和迷恋。

楹联是中国特有的文学形式，言简意深，对仗工整，平仄协调，在形式上它极富对称之美，在内容上则"阴阳合而变化起"，或诗词或曲赋，结合汉字的声形特征，将中国传统文化的精髓表现得淋漓尽致。

楹联又称对子、对联，在呈现形式上，或写在纸上、布绢上，或刻在木板、柱头上。相传五代后蜀孟昶，在他寝室门桃符板上的题词——"新年纳余庆，嘉节号长春"，是楹联的开端之作。而经过长期的发展与演变，楹联早已深入中国社会生活的各个领域，尤其旅游胜地、亭台楼阁、墓祠庙宇、会馆建筑等，是楹联集大成处。它们既是这些场所的山水吟唱、文史典故介绍，楹联本身的装饰作用，也为建筑物增色不少。

与其他楹联有所不同的是，洛带会馆中为数众多的楹联，大都是述说移民思乡之情、同乡之谊、努力创业、歌颂先贤、勉励后代的。譬如湖广会馆有联——"看大江东去穿洞庭出鄂渚水天同一色纪功原是故乡梦，策匹马西来寻石纽问涂山圣迹几千里望古应知明月远"表达了湖广移民对故土的思恋和先贤的崇敬；广东会馆有联——"庙堂经过劫灰年宝相依然重镇曹溪钟鼓，华简俱成桑梓地乡音无改新增天府冠裳"，表达了粤籍客家人心念先贤，以他乡为故乡，勇于开拓的精神。

可见楹联对洛带会馆而言，既是与会馆建筑相宜得章的审美体现，更是解读移民心灵密码的钥匙。

CHAPTER 7
Couplets Hung on the Columns of Guild Halls
The Keys to the Soul of Migrants

All the major guild halls located in Luodai adopted a symmetrical layout with an axle wire—hall for worship, meeting hall, exquisite or grand stage, alignment of architectures with two or multiple clusters inside the axle, wing-room for guests along the axle wire.

In addition, Huguang Guild Hall, Jiangxi Guild Hall, and Guangdong Guild Hall were built up in Luodai originally all facing the south, which is the result of their purest idea that according to Laozi, a sage, "all things leave behind them belong to yin (negative, dark, and feminine), and embrace yang (positive, bright, and masculine), with a new harmonious article formed through the collision between yin and yang", as well as the idea that the guild halls should face their hometowns. In Chinese people's traditional geomancy notion, the north represents water and is the yin, while the south represents fire and is the yang. Therefore, most Chinese architectures face the south and embrace the yang.

The binary idea of yin and yang is not only the basic world outlook of Chinese people, but also the basis of Chinese culture. The idea of yin and yang that "the world consists of left and right as well as yin and yang" is deep-rooted in Chinese people's mind. As it is reflected as a kind of national psychology, one of the most important features is the adherence to "beauty of symmetry".

Couplet is a unique literary form in China, with deep meaning in brief language, neat parallelism, harmonious level and oblique tones, symmetrical beauty in form, and "yin and yang change alternatively and successively" in content. It is poetry of different dynasties and in various forms, which fully express the essence of traditional Chinese culture through making use of the features of Chinese characters.

Couplet is also called a pair of antithetical phrases or antithetical couplet. It can be written on paper, cloth and tough silk, and carved on board or chapiters. It is said that Meng Chang, the last emperor of later Shu Kingdom during the Five Dynasties, wrote an inscription containing charms against evil on the peach wood of his bedchamber–"the kindness and charity left by our ancestors are enjoyed in a new year; the joyous festival also indicates a long spring", which was the beginning of the couplet. After long-term development and evolvement, couplets are deeply rooted in every field of China's social life, especially in tourist attractions, pavilions, terraces, towers, tombs, ancestral halls, temples, guild halls, etc., which comprehensively express couplets. They highly praise mountain and water landscape and introduce cultural and historical allusions of those venues. As a kind of decoration, couplets make architectures more attractive.

Unlike other couplets, the numerous couplets of Luodai Guild Halls mostly are expressing immigrants' homesickness, the friendship among people from the same home towns, and immigrants' hard work, singing the praise of sages, and encourage descendants. A pair of couplets of Huguang Guild Hall say–"the mighty Yangtze River flows eastward through Dongting Lake via Echu, with the waters and skies merging in one color, but such an achievement statement was only a dream about hometown; we spur a horse westward to seek Shiniu and ask how far away from the holy land Tushan, and we knew that the bright moon is far in spite of our outlook towards our source", expressing that immigrants from Huguang suffered homesickness and held respect for sages there; Guangdong Guild Hall has a couplet–"although imperial court suffered flame of war but the appearance of Buddhist is still, which will make the bell drum of Caoxi Temple calm again; Huayang and Jianyang both become their hometowns, without any change in their dialect, which added new charms to Chengdu", expressing that Hakka people from Guangdong kept thinking of sages and took other places as their hometown thanks to their pioneering spirit.

We can learn that for guild halls in Luodai, couplet plays the role as not only an appreciation of beauty, but also a key to immigrants' souls.

前殿明间檐柱联——
与二帝为三精一危微执厥中而膺历数
冠百王之首荡平正直握皇极以叙彝伦

前殿次间檐柱联——
继尧舜传心先汤文立极；
拯人民饥溺辅天地平成

后殿明间前檐柱联——传子即传贤,天下为公同尧舜;治民先治水,山川永奠泽重湖湘。

洛带湖广会馆楹联节选

Excerpts from Couplets Hung on the Columns of Huguang Guild Hall in Luodai

洛带江西会馆楹联节选

Excerpts from Couplets Hung on the Columns of Jiangxi Guild Hall in Luodai

大门明间楹联——
日出东山看落带楼台四面桃花映绿水闻鸡鸣犬吠牛马喧此地恰似武陵胜地；
客来南海兴江西会馆八方贤达话青茶喜果肥花密稻麦香这里依稀蓬莱仙家。

大门次间楹联——福地仙栖群英多奉献；明时人乐三蜀美乡帮；

后殿后檐柱楹联——创业东山后嗣无忘祖德；守成西蜀先人喜作孙谋。

CHAPTER 7 会馆楹联　解读移民心灵密码的钥匙

135

前殿次间檐柱楹联——惨淡经营，庙貌独怜供一炬；努力缔造，神庥共祝保千秋。

中殿明间檐柱楹联——庙堂经过劫灰年，宝相依然，重振曹溪钟鼓；华简俱成桑梓地，乡音无改，新增天府冠裳。

后殿底楼明间前檐柱楹联——云水苍茫,异地久栖巴子国;乡关迢递,归舟欲上粤王台。

洛带广东会馆楹联节选

Excerpts from Couplets Hung on the Columns of Guangdong Guild Hall in Luodai

洛带会馆中为数众多的楹联，大都是述说移民思乡之情、同乡之谊、努力创业、歌颂先贤、勉励后代的。楹联对洛带会馆而言，既是与会馆建筑相宜得章的审美体现，更是解读移民心灵密码的钥匙。

The numerous couplets of guild halls in Luodai express migrants' homesickness, the friendship among people from the same home towns, and migrants' hard work, sing the praise of sages, and encourage descendants. For guild halls in Luodai, couplets function not only as an appreciation of beauty, but also as a key to migrants' souls.

Opera Stages in Guild Halls

第八章
会馆戏楼

初晴天气，观剧人伙，万头攒簇，庙坝为满 ——（清）丁治棠《晋省记》

会馆戏楼 川剧五腔共和的摇篮

第八章

　　祭祀是会馆的一项重要功能，因为要祭祀神祇，就要唱戏酬神；因为要唱戏酬神，各籍移民会馆都建有别具一格的戏台。在洛带，湖广会馆、广东会馆、川北会馆，都将其戏台设置在门楼的里侧，因为这些门楼都对应着一个开放而疏阔的庭院，更对着会馆大殿里祀奉的诸神。而江西会馆除了门庭之外巍峨的万年台，尚在内部庭院设置了一座精巧的小戏台。

　　通过酬神的戏曲演出，徐真人、粤王、大禹等神化的先贤和世俗的移民，一起度过一段快乐的时光，那真是酬神娱人两不误。到后来，凡节庆之时，会馆会期，不分籍贯彼此，人们都会在会馆里济济一堂，共赏戏曲之乐。可以想见，其时的洛带，从上街到下街，各家会馆是你方唱罢我登场，四时热闹不断。

　　事实上，受原乡地域文化的影响，入川移民对戏曲的爱好也因地域而别。一般说来，江、浙、闽、粤之人好昆腔；赣、湘、鄂之人好高腔、胡琴；陕、甘之人爱弹戏，本土的乡人喜灯戏，在苦恋的故乡声腔中，移民们一度保持着对故乡戏曲的热爱。

　　不过，随着时间的推移，移民后代已不如父辈那样醉心乡音了。作为这片土地新的主人，他们对时下流行的声腔和名角更有广泛的兴趣，所以时有《锦城竹枝词》说，"见说高腔有苟莲，万头攒动万家传"。与此同时，各地名伶也彼此观摩，切磋技艺，并苦练唱功、做功，以"兼擅文武"、"昆乱不挡"、一专多能为荣。于是一种荟萃昆、高、胡、弹、灯的戏班便应运而生，并在民间年节和会馆会期的迎神、报赛、宴会、酬宾中受到极大欢迎，所谓"鱼龙曼衍，百戏杂陈。士民走观，充衢溢巷"。

　　那时的成都，尚无戏院之说，各种演出活动多在会馆的戏台举行。清光绪年间丁治棠所著《晋省记》曾详细记载了成都会馆观戏盛况——"初晴天气，观剧人伙，万头攒簇，庙坝为满。茶担木凳，无隙可坐。是时，人涌如潮，吹哨喝彩者，应声四起。女坐边凳，随挤而倾，诸茶担为之震摇"。正是如此的会馆戏剧之乐，"五腔共和"的川剧表演形式最终在辛亥革命后成形，这一点，会馆功不可没，会馆戏台可谓川剧诞生的摇篮。

CHAPTER 8
Opera Stages in Guild Halls
Where 5 Common Systematic Tunes of Sichuan Opera Blend

Offering sacrifices to gods was an important function of guild halls. Operas were sung for gods to accompany the sacrifices offering. In order to sing opera, a unique stage was built in each immigrant guild hall. In Luodai, the Huguang Guild Hall, the Canton Guild Hall and the North Sichuan Guild Hall have a stage inside their gatehouses. These gatehouses are not only facing a wide open courtyard but also the gods enshrined in the lobby of the guild halls. On the other hand, the Jiangxi Guild Hall, located next to the towering Wannian Stage outside its gatehouse, has an elaborate internal courtyard.

Opera performances entertained secular immigrants as well as Immortal Xu, Guangdong King, Dayu and other deities. At every festival and guild hall meeting, regardless of origin, people gathered together in guild halls to enjoy operas. It was very boisterous in every guild hall on each street in Luodai.

In fact, influenced by their hometown culture, immigrants in different regions of Sichuan had varied preferences for operas. Generally, people from Jiangsu, Zhejiang, Fujian and Guangdong preferred Kun tune while people from Jiangxi, Hunan and Hubei favored high-pitched tune and huqin. People from Shaanxi and Gansu Tan preferred dramas; whereas, native countrymen were fond of Lantern dramas. With keen longing for tunes from their hometowns, immigrants maintained a passion for hometown operas for a long period of time.

However, descendants of these immigrants were less bonded with tunes from their origins. As new owners of this land, they were more interested in popular tunes and performers. There was a ballad that roughly goes: "If a high-pitched tune is staged, thousands of people will go to watch it and it will be sung by hundreds of households". At the same time, famous opera performers from various places also learned from each other and tried hard to improve their singing and performing skills. They took pride in versatility and multi-skill on drama acting. Therefore troupes that were capable of Kun tune, high-pitched tune, huqin, Tan dramas and Lantern dramas came into being and were well received by people during festivals and various guild hall events, like gods reception, gods worshiping, banquets and guests entertainment. Just as an ancient person describes, various dramas were blended and all streets and alleys were full of people watching them.

There was no theatre in Chengdu at that time. Most performances were given in guild halls. Ding Zhitang described a spectacular opera-watching scene in Chengdu's guild hall in Emperor Guangxu's reign of the Qing Dynasty in the book Jinshengji. "It's a hard-earned day. The opera began. Thousands of spectators crowded the courtyard of the guild hall. Crowded wooden benches and shoulder poles for carrying tea and left no space for people to sit. People were excited, and were moving like waves. Whistles and cheers were heard everywhere. Stools were even falling and the shoulder poles were shaken".

It was these operas in guild halls that facilitated the blending of five common systematic tunes of Sichuan Opera and established it after the Revolution of 1911. Guild halls contributed a lot to this and can be described as the cradle of Sichuan Opera.

移民会馆戏台的本意是酬神，并通过酬神戏曲的乡音乡情，崇祀神明，敦睦乡谊。而随着移民社会的发育和交融，各籍移民会馆的地方戏曲表演也就愈加娱人化了，且互为借鉴，逐渐融汇，热热闹闹地向川剧"五腔共和"的道路迈进。与此同时，出于民众对戏曲的热爱，会馆戏台的建造更加讲究，更为精美。

Theatrical stages in migrant's guild halls were initially intended for god worship and promote friendly relations among hometown fellows through hometown operas. With the development and blending of migrant society, opera performances in various migrant guild halls became more entertaining, learned from each other and gradually blended, livelily marching to the destination where 5 common systematic tunes of Sichuan Opera blend. Meanwhile, because of people's zealous love for operas, theatrical stages in guild halls were built to be more exquisite with more care.

锣鼓丝竹之音响起，各方英雄豪杰、才子佳人又在洛带会馆古老的戏台上粉墨登场。过去，同乡之人在此共叙乡情，共赏戏曲之乐；今天，戏台下满座的是醉迷在浓郁客家文化中的八方宾客。

To the sounds of gongs and drums, heroes, gifted scholars and beautiful ladies in dramas appeared on the ancient stages in the guild halls in Luodai. In the past, hometown fellows talked about nostalgia and enjoyed operas here. Today a full house of guests from all over the world are fascinated by attractive and profound Hakka culture.

得益于会馆戏曲的陶冶，洛带一地"玩友"众多。每月十五，他们都会在洛带会馆的戏台上会聚。随着一阵鼓钹之音紧密奏响又嘎然而止，《贵妃醉酒》、《穆桂英打雁》等川剧唱段高亢地唱起。有没有观众不要紧，全凭自己过瘾就好。寻着会馆飞檐飘荡而去的锣鼓之音和声声唱腔，就这样不经意间热闹了古镇一段悠扬的时光。

　　Thanks to the influence of operas in guild halls, there are many amateur opera performers in Luodai. On the fifteenth day of every lunar month, they gather on the theatrical stages of guild halls in Luodai. With a burst of drums and cymbals, "the Drunken Beauty", "The Wild Goose Hunting by Mu Guiying" and other Sichuan Operas are shown. It does not matter whether there is audience or not, they just want to enjoy themselves. The sound of gongs, drums and singing gets far away along the overhanging eaves of guild halls and inadvertently livens up the ancient town.

乡音乡情，唱不尽的先贤恩德和故土眷念；耕读为本，道不尽的礼乐文明和语重心长的谆谆教诲……翠竹掩映的洛带广东会馆戏楼，悠扬婉转的客家山歌，描绘出一卷风情洋洋的岭南风俗画卷。

Hometown operas are full of nostalgia as well as kindness from sages. Oriented by farming and learning, operas eulogize endless civilization of rites and music and preach teachings earnestly... the theatrical stage in the bamboo-hidden Canton guild hall in Luodai and the melody of the Hakka folk songs depict a picture of customs in the south of the Five Ridges.

CHAPTER 8 会馆戏楼 川剧五腔共和的摇篮

　　过去，洛带各家会馆每年都会如期举行庙会，除各籍移民所崇信先贤乡神的祭祀庙会外，各家会馆在诸如春节这样的重要节气期间，也有盛大的庙会活动举行。会馆庙会期间，不仅有戏剧表演，更有傩舞或者体现原乡风俗的巡游活动，娱神娱人，热闹非凡。会馆巡游，实则体现了移民社会初期，共克时艰的移民们渴望虚拟戏剧生活的一种心理暗示；其上承古韵、具有鲜活生活基础的表演方式，是川剧成型与发展取之不竭的艺术创造源泉。

　　浩浩荡荡的会馆巡游队伍，幡旗飘飘地行走在会馆荟萃的洛带古镇上，别开生面，独具风情。

　　In the past, people gathered in the guild halls in Luodai on schedule every year. In addition to gatherings for worshiping sages and gods that all migrants believe in, a grand temple fair was held in each guild hall during the Spring Festival and other important solar terms. During gatherings in the guild halls, there were operas, Nuo dance, or parade that reflects customs in migrants' hometowns, which were extraordinarily bustling and entertained both gods and human beings. The parade of guild halls actually reflected the migrants' longing for the life in virtual dramas at early times when they needed to overcome many difficulties. The fine archaic rhyme and the fresh and alive life-based way of performances were an inexhaustible source of artistic creation in forming and developing Sichuan Opera.

　　Waving flags, the large parade of guild halls marched in Luodai Ancient Town, where many guild halls gather innovatively and uniquely.

图书在版编目（CIP）数据

洛带会馆 客地原乡：汉、英 / 张利频，曾列主编. -- 成都：四川美术出版社，2013.10
ISBN 978-7-5410-5714-4

Ⅰ.①洛… Ⅱ.①曾… Ⅲ.①会馆公所－古建筑－介绍－成都市－汉、英 Ⅳ.①TU-092.2

中国版本图书馆CIP数据核字(2013)第249308号

主　　编：	张利频　曾　列
执行主编：	张　珂
副 主 编：	郭政权　万玉林
编委会：	张利频　曾　列　郭政权　张　珂　万玉林
	樊　琴　袁云基　黄　蓉　钟国勇　张世凯
	杨　军　叶云君　张亦龙　郑晓娟

洛带会馆 客地原乡
LUO DAI HUI GUAN　KE DI YUAN XIANG

张利频　曾列　主编

出 品 人：	马晓峰
科学顾问：	王小灵
责任编辑：	陈　晶
作　　者：	余茂智
书名题写：	郭政权
设　　计：	曾　云
英文翻译：	四川语言桥信息技术有限公司
责任校对：	曾品艳　万　程
出版发行：	四川出版集团　四川美术出版社
地　　址：	四川省成都市三洞桥路12号（610031）
制　　作：	成都墨道文化传播有限公司
印　　刷：	四川新华彩色印务有限公司
承　　制：	成都九泰文化传播有限公司
成品尺寸：	210mm×285mm
印　　张：	9.5
字　　数：	400千字
图　　幅：	150幅
版　　次：	2013年11月第1版
印　　次：	2013年11月第1次印刷
书　　号：	ISBN 978-7-5410-5714-4
定　　价：	328.00元

■ 著作权所有　违者必究
■ 本书若出现印装质量问题，请与工厂联系调换